Profiles of Space Scientists: Pioneering Minds in Cosmic Exploration

Shah Rukh

Published by Shah Rukh, 2024.

PROFILES OF SPACE SCIENTISTS: PIONEERING MINDS IN COSMIC EXPLORATION

First edition. June 2, 2024.

Written by Shah Rukh.

Table of Contents

Prologue

The night sky, with its myriad twinkling stars, has always been a source of wonder and curiosity for humankind. From the earliest civilizations that charted the heavens to navigate and mark the passage of time, to modern astronomers who peer deep into the cosmos with powerful telescopes, our fascination with space has driven us to explore beyond the confines of our planet. "Profiles of Space Scientists: Pioneering Minds in Cosmic Exploration" is a journey through the lives and achievements of the remarkable individuals who have expanded our understanding of the universe.

In this book, we delve into the stories of fifty extraordinary scientists, each a luminary in their field, whose work has shaped the course of astronomical and space science. These pioneers, spanning centuries and disciplines, have laid the groundwork for our current knowledge and inspired generations of scientists and explorers.

Our journey begins with Galileo Galilei, whose telescopic observations challenged the geocentric worldview and paved the way for modern astronomy. We then follow the paths of Johannes Kepler and Christiaan Huygens, who elucidated the laws of planetary motion and the nature of light, respectively. Each subsequent chapter introduces us to new visionaries: from Isaac Newton's revolutionary laws of motion and gravity to James Clerk Maxwell's electromagnetic theory, and from Konstantin Tsiolkovsky's dreams of space travel to Wernher von Braun's realization of those dreams with the development of rocket technology.

We explore the contributions of astrophysicists like Edwin Hubble, who expanded our universe, and cosmologists like Georges Lemaître, who proposed the Big Bang theory. The book also highlights the work of engineers and astronauts, including Sergei Korolev, the mastermind behind Soviet space achievements, and Neil Armstrong, the first human to set foot on the moon.

The inclusion of often-overlooked figures such as Annie Jump Cannon and Henrietta Swan Leavitt, whose work on stellar classification and luminosity fundamentally altered our understanding of the stars, ensures a comprehensive look at the diverse minds that have propelled our cosmic exploration forward. The book also honors the contributions of women and minorities in the field, like Katherine Johnson and Mae Jemison, whose determination and brilliance broke barriers and opened doors for future generations.

As we navigate through these chapters, we witness the evolution of space science from theoretical musings to tangible achievements. We see the shift from ground-based observations to space-based explorations, from the first artificial satellites to the International Space Station, and from the conceptualization of black holes to the detection of gravitational waves.

"Profiles of Space Scientists" is not just a collection of biographies; it is a tribute to the relentless curiosity, ingenuity, and perseverance of these trailblazers. Their stories remind us that the quest for knowledge is a journey fraught with challenges, yet it is through overcoming these challenges that we achieve our greatest triumphs.

In an era where space exploration is becoming increasingly collaborative and global, this book serves as a testament to the power of human spirit and intellect. It celebrates the scientists who have illuminated the darkness of the unknown and continue to inspire us to reach for the stars. As we look to the future of space exploration, may the legacies of these pioneering minds guide and inspire the next generation of explorers, ensuring that our journey through the cosmos continues unabated.

Chapter 1: Galileo Galilei

Galileo Galilei, born on February 15, 1564, in Pisa, Italy, is often heralded as the "father of modern observational astronomy," the "father of modern physics," and the "father of modern science." His contributions to the scientific revolution of the 17th century were foundational, particularly in the realms of astronomy, physics, and the scientific method.

Galileo was the eldest of six children in a family of modest means. His father, Vincenzo Galilei, was a renowned lutenist and music theorist who played a significant role in shaping Galileo's early intellectual development. Despite financial constraints, Vincenzo ensured that his son received a well-rounded education. Initially, Galileo was sent to the University of Pisa to study medicine, but he soon developed a strong interest in mathematics and natural philosophy, largely influenced by the lectures of Ostilio Ricci, a mathematician in the court of the Grand Duke of Tuscany.

In 1589, Galileo secured a position as a mathematics professor at the University of Pisa. During this period, he began to challenge the Aristotelian physics that dominated the academic landscape. One of his most famous experiments, allegedly conducted from the Leaning Tower of Pisa, involved dropping two spheres of different masses to demonstrate that their time of descent was independent of their mass. This experiment directly contradicted Aristotle's theory of gravity and laid the groundwork for Galileo's later work on motion.

In 1592, Galileo moved to the University of Padua, where he taught geometry, mechanics, and astronomy until 1610. His years in Padua were highly productive. It was here that he invented the thermoscope, a forerunner of the thermometer, and devised a geometric and military compass used for practical applications such as gunnery and surveying. These inventions garnered him considerable fame and financial gain, allowing him to focus more on his research.

Galileo's interest in astronomy was piqued in 1609 when he learned of the invention of the telescope in the Netherlands. He swiftly built his own improved version, with which he made several groundbreaking discoveries. In 1610, he published "Sidereus Nuncius" (Starry Messenger), a short treatise that detailed his observations of the Moon's surface, which he found to be mountainous and pocked with craters, contradicting the Aristotelian belief in a perfect, smooth lunar surface. He also discovered four moons orbiting Jupiter, which he named the Medicean Stars in honor of his patrons, the Medici family. This was a significant blow to the geocentric model of the cosmos, which held that all celestial bodies revolved around the Earth.

Galileo's telescopic observations did not stop there. He observed the phases of Venus, which provided strong evidence for the Copernican heliocentric model, where planets orbit the Sun. He also noted the existence of sunspots and the rotation of the Sun on its axis, further challenging the notion of unchanging heavens.

Despite his monumental contributions to science, Galileo's support for the heliocentric model brought him into conflict with the Roman Catholic Church. The Copernican system was controversial because it seemed to contradict certain passages of the Bible and the long-held geocentric model endorsed by the Church. In 1616, the Inquisition condemned heliocentrism as "formally heretical," and Galileo was warned to abandon his support for it. Although he complied publicly, he continued his research in private.

Galileo's most comprehensive work on heliocentrism, "Dialogue Concerning the Two Chief World Systems," was published in 1632. This book presented a dialogue between proponents of the geocentric and heliocentric systems, ultimately favoring the latter. The work was perceived as an open endorsement of Copernicanism, leading to Galileo's trial by the Inquisition in 1633. He was found "vehemently suspect of heresy" and forced to recant his views. He spent the remainder of his life under house arrest.

Even during his confinement, Galileo continued his scientific work. He wrote "Discourses and Mathematical Demonstrations Relating to Two New Sciences," which summarized much of his earlier work on kinematics and strength of materials. This book, published in 1638, is considered his finest work and laid the foundation for classical mechanics.

Galileo's contributions extended beyond his discoveries. He was a pioneering advocate for the scientific method, emphasizing observation, experimentation, and mathematical analysis as the keys to understanding the natural world. His insistence on questioning established doctrines and relying on empirical evidence was revolutionary and set the stage for future scientific inquiry.

Galileo passed away on January 8, 1642, in Arcetri, near Florence. His death did not mark the end of his influence. His ideas continued to inspire scientists, and his methods became integral to the development of modern science. Galileo's legacy is profound; he transformed our understanding of the universe and laid the groundwork for future scientific advancements. His life's work not only challenged the prevailing views of his time but also established principles that remain at the heart of scientific investigation today. His courage in facing opposition and his relentless pursuit of knowledge exemplify the spirit of scientific inquiry, making him a towering figure in the history of science.

Chapter 2: Johannes Kepler

Johannes Kepler, born on December 27, 1571, in the Free Imperial City of Weil der Stadt, located in present-day Germany, was a key figure in the scientific revolution and is most renowned for his laws of planetary motion. His contributions to astronomy, mathematics, and optics laid crucial foundations for modern science.

Kepler's early life was marked by hardship. His father, Heinrich Kepler, was a mercenary who abandoned the family when Johannes was young. His mother, Katharina Guldenmann, was an herbalist who later faced accusations of witchcraft. Despite these adversities, Kepler displayed an early aptitude for mathematics and astronomy. He attended the University of Tübingen, initially studying theology. At Tübingen, he was introduced to Copernican theory, which posited that the planets orbit the Sun. This heliocentric model profoundly influenced Kepler's thinking.

In 1594, Kepler became a mathematics teacher at a Protestant school in Graz, Austria. Here, he began his work on the Mysterium Cosmographicum (The Cosmographic Mystery), his first major astronomical work, published in 1596. Kepler sought to explain the distances of the planets from the Sun by inscribing and circumscribing spheres and regular polyhedra, demonstrating a deep belief in the geometric nature of the cosmos. While some of his geometric assumptions were later proven incorrect, this work laid the groundwork for his future discoveries.

Kepler's astronomical career took a significant turn in 1600 when he met Tycho Brahe, a prominent Danish astronomer, in Prague. Brahe had accumulated vast amounts of precise astronomical data, especially concerning the motion of Mars. Kepler, possessing a brilliant mathematical mind, was tasked with analyzing this data. Upon Brahe's death in 1601, Kepler succeeded him as the Imperial Mathematician to Emperor Rudolf II, gaining access to Brahe's extensive observations.

Kepler's meticulous analysis of Brahe's data led to the formulation of his three laws of planetary motion, which revolutionized our understanding of the solar system. The first law, the Law of Ellipses, stated that planets orbit the Sun in elliptical paths, with the Sun at one focus. This law, published in 1609 in his work Astronomia Nova (New Astronomy), shattered the long-held belief in circular orbits, a notion stemming from Aristotelian and Ptolemaic astronomy. The second law, the Law of Equal Areas, also published in Astronomia Nova, declared that a line segment joining a planet and the Sun sweeps out equal areas during equal intervals of time. This implied that planets move faster when they are closer to the Sun and slower when they are farther away, indicating a variable speed in their orbits.

Kepler's third law, the Harmonic Law, was introduced in his later work Harmonices Mundi (Harmony of the Worlds), published in 1619. This law stated that the square of a planet's orbital period is proportional to the cube of the semi-major axis of its orbit. Mathematically, this is expressed as $T^2 \propto r^3$, where T is the orbital period, and r is the average distance from the Sun. This relationship provided a precise mathematical description of the solar system's structure and was crucial for later developments in celestial mechanics.

In addition to his laws of planetary motion, Kepler made significant contributions to the field of optics. His work Astronomiae Pars Optica (The Optical Part of Astronomy), published in 1604, laid the foundation for modern optics. He studied the behavior of light, including reflection, refraction, and the formation of images in the human eye. He also explained the principles of the camera obscura, which was an early precursor to the modern camera. In 1611, he published Dioptrice, where he elaborated on the properties of lenses and improved the design of telescopes, significantly enhancing their magnifying power.

Kepler's contributions were not limited to theoretical work; he was also involved in practical astronomy. He constructed the Rudolphine

Tables, published in 1627, which were the most accurate astronomical tables of their time. These tables were based on the precise observational data of Tycho Brahe and Kepler's laws of planetary motion, allowing for improved predictions of planetary positions and eclipses.

Despite his groundbreaking achievements, Kepler's life was fraught with personal and professional challenges. The political and religious turmoil of the Thirty Years' War affected his career and personal life. As a Protestant, he faced religious discrimination and was forced to relocate multiple times. His mother's trial for witchcraft, though ultimately resulting in her acquittal, added to his distress.

Kepler's later years were marked by continued scientific inquiry and publication. His work Somnium (The Dream), published posthumously, is considered one of the first works of science fiction, describing a journey to the Moon and speculating about lunar inhabitants. This work reflected Kepler's imaginative and speculative approach to scientific questions.

Kepler died on November 15, 1630, in the city of Regensburg, while traveling to collect an outstanding salary. He was buried in the local churchyard, but the grave was destroyed during the subsequent Swedish invasion in the Thirty Years' War. Despite these hardships, Kepler's legacy endured. His laws of planetary motion were pivotal for Isaac Newton's formulation of the law of universal gravitation. Newton famously acknowledged Kepler's influence, stating, "If I have seen further, it is by standing on the shoulders of giants."

Kepler's work exemplifies the transition from classical to modern science. His meticulous analysis of empirical data, combined with his mathematical prowess, demonstrated the power of observation and theory in unison. His laws of planetary motion remain fundamental to our understanding of the cosmos, and his contributions to optics and mathematics continue to resonate in contemporary science. Kepler's life and work reflect a profound commitment to uncovering the mysteries of the natural world, driven by a deep-seated belief in the harmony and order of the universe. His enduring legacy is a testament to the transformative power of scientific inquiry and the relentless pursuit of knowledge.

Chapter 3: Christiaan Huygens

Christiaan Huygens, born on April 14, 1629, in The Hague, Netherlands, was a towering figure in the scientific revolution of the 17th century. His contributions spanned a wide array of fields, including mathematics, physics, astronomy, and horology, and his work laid the groundwork for many developments in modern science. Huygens' life was one of intense intellectual pursuit, marked by groundbreaking discoveries and innovations.

Huygens was born into a wealthy and influential family. His father, Constantijn Huygens, was a prominent diplomat, poet, and advisor to the House of Orange. This background provided Christiaan with a robust education and early exposure to the leading intellectual figures of his time. From a young age, Huygens showed an exceptional aptitude for mathematics and science, which his father nurtured through private tutors and a rich intellectual environment.

At the age of sixteen, Huygens enrolled at the University of Leiden, where he studied law and mathematics. He was particularly influenced by the lectures of Frans van Schooten, a mathematician who introduced him to the works of René Descartes and other prominent thinkers. Huygens continued his studies at the newly founded Orange College in Breda, further honing his mathematical skills and developing an interest in natural philosophy.

Huygens' early work in mathematics was impressive. He made significant contributions to the field of probability theory, notably through his 1657 publication "De ratiociniis in ludo aleae" (On Reasoning in Games of Chance), which was one of the first books to systematically analyze probability. His insights into games of chance were not only mathematically rigorous but also practical, influencing gambling practices and insurance calculations.

In 1655, Huygens turned his attention to astronomy, driven by the advancements in telescope technology pioneered by Galileo Galilei.

Using a telescope he constructed himself, Huygens made several significant astronomical discoveries. His most famous discovery came in March 1655 when he identified Titan, the largest moon of Saturn. This discovery was monumental as Titan was the first moon of Saturn to be observed, expanding our understanding of the solar system.

Huygens continued his observations of Saturn and, in 1659, published "Systema Saturnium" (The System of Saturn). In this work, he correctly described the nature of Saturn's rings, which had puzzled astronomers since Galileo's initial observations. Huygens proposed that the rings were composed of numerous small particles orbiting the planet, a hypothesis that, while not entirely accurate, was closer to the truth than previous theories.

In addition to his astronomical achievements, Huygens made significant contributions to optics. He developed a new method for grinding and polishing lenses, which greatly improved the quality and magnification of telescopes. His work on the wave theory of light, published in his 1690 treatise "Traité de la lumière" (Treatise on Light), was revolutionary. Huygens proposed that light traveled in waves, which explained various optical phenomena such as reflection, refraction, and diffraction. His wave theory of light was initially overshadowed by Isaac Newton's corpuscular theory but later became a cornerstone of modern optics.

Huygens' contributions to mechanics and physics were equally groundbreaking. He formulated the laws of motion for bodies in a collision, laying the foundation for the principle of conservation of momentum. In his 1673 work "Horologium Oscillatorium" (The Pendulum Clock), Huygens presented his mathematical analysis of pendulum motion. He derived the law of centrifugal force for uniform circular motion and described the concept of centripetal force. This work was crucial in the development of classical mechanics and had a profound influence on Isaac Newton's formulation of his laws of motion.

One of Huygens' most practical inventions was the pendulum clock, which he patented in 1657. Prior to Huygens' invention, mechanical clocks were notoriously inaccurate. The introduction of the pendulum as a timekeeping element vastly improved the accuracy of clocks, reducing the margin of error to within seconds per day. This innovation had a significant impact on navigation, allowing sailors to determine longitude more precisely and contributing to safer and more efficient sea travel.

Huygens also explored the nature of sound and its propagation. He was one of the first to accurately describe the phenomenon of beats, which occur when two sound waves of slightly different frequencies interfere with each other. His studies in acoustics further demonstrated his versatile scientific curiosity and his ability to apply mathematical principles to understand natural phenomena.

In addition to his scientific endeavors, Huygens was deeply involved in the intellectual community of his time. He corresponded with many of the leading scientists and philosophers, including René Descartes, Isaac Newton, and Gottfried Wilhelm Leibniz. His extensive correspondence provides valuable insights into the scientific discourse of the 17th century and highlights Huygens' role as a central figure in the scientific revolution.

Huygens' later years were marked by continued scientific inquiry and innovation. He moved to Paris in 1666, where he became a founding member of the French Academy of Sciences. During his time in Paris, Huygens continued his research on optics, mechanics, and astronomy. However, his health began to decline, and in 1681, he returned to the Netherlands. Despite his ailments, he continued to work on scientific problems until his death on July 8, 1695.

Christiaan Huygens' legacy is profound and far-reaching. His discoveries and inventions laid the groundwork for many fields of science and technology. The wave theory of light, the laws of motion for colliding bodies, the accurate timekeeping provided by the pendulum clock, and his contributions to probability theory are all testaments to his genius. Huygens' work exemplifies the spirit of the scientific revolution, characterized by rigorous mathematical analysis, empirical observation, and a relentless pursuit of knowledge. His contributions continue to influence modern science and technology, underscoring his lasting impact on the world.

Chapter 4: Giovanni Cassini

Giovanni Domenico Cassini, also known as Jean-Dominique Cassini in France, was an Italian-born astronomer, engineer, and astrologer whose work significantly advanced the fields of astronomy and planetary science. Born on June 8, 1625, in Perinaldo, Republic of Genoa (now part of Italy), Cassini's career spanned a period of intense astronomical discovery and innovation. He is best known for his discoveries related to the planet Saturn, including its moons and the division of its rings, as well as for his pioneering work in creating one of the first comprehensive maps of the moon.

Cassini's early education was marked by a strong focus on mathematics and astronomy. He was initially educated at a Jesuit school where he developed an early interest in astrology and mathematics. His talents in these fields were quickly recognized, and by the age of 25, he had secured a position at the Panzano Observatory, located in Bologna, Italy. At Panzano, under the mentorship of the mathematician and astronomer Giovanni Riccioli, Cassini honed his observational skills and began making significant contributions to the field of astronomy.

One of Cassini's early achievements at Panzano was his work on the meridian line, a precursor to the modern concept of a prime meridian. This involved accurately determining the longitude and latitude of various locations, which was crucial for navigation and cartography. His meticulous observations and calculations laid the groundwork for more precise geographic and astronomical measurements.

In 1650, Cassini was appointed as a professor of astronomy at the University of Bologna, a position he held for over two decades. During this period, he conducted extensive observations of the planets and stars, improving the accuracy of planetary tables and refining the methods used to calculate their positions. His work was characterized by a rigorous approach to observational astronomy and a deep commitment to advancing the precision of astronomical measurements.

Cassini's reputation as a leading astronomer reached across Europe, attracting the attention of the French King Louis XIV. In 1669, Cassini was invited to France to join the newly established Paris Observatory, which had been founded by the French Academy of Sciences. Accepting the invitation, Cassini moved to Paris and became a French citizen, adopting the name Jean-Dominique Cassini. At the Paris Observatory, Cassini's work flourished, and he made some of his most famous discoveries.

One of Cassini's most notable achievements was his study of the planet Saturn. Through careful observation using telescopes that were among the most advanced of his time, Cassini discovered four of Saturn's moons: Iapetus in 1671, Rhea in 1672, Tethys in 1684, and Dione in 1684. His discovery of these moons significantly expanded our understanding of the Saturnian system and the diversity of celestial bodies within our solar system.

In addition to discovering Saturn's moons, Cassini is best known for identifying a division within Saturn's rings, which is now known as the Cassini Division. This dark gap between the planet's A and B rings, discovered in 1675, provided crucial evidence that Saturn's rings were not a solid, continuous structure but rather composed of countless small particles orbiting the planet. This finding was instrumental in shaping our understanding of the dynamics and composition of planetary rings.

Cassini's work extended beyond the study of Saturn. He made significant contributions to the field of astrometry, the precise measurement of the positions and movements of stars and planets. His meticulous observations and calculations led to more accurate planetary tables and improved navigation techniques. Cassini's work on the moons of Jupiter, particularly his observations of their eclipses, provided essential data for determining the longitude of various locations on Earth, which was crucial for navigation and mapping.

One of Cassini's major projects at the Paris Observatory was the creation of a comprehensive map of the moon. Working with the Italian-French astronomer Jean Richer, Cassini used a method called micrometry to measure the positions and sizes of lunar features with unprecedented accuracy. The resulting map, published in 1679, was one of the most detailed and accurate lunar maps of its time, contributing to the field of selenography, the study of the surface and physical features of the moon.

Cassini's contributions to astronomy were not limited to his observational work. He was also an innovator in the design and use of telescopes and other astronomical instruments. He played a key role in the development of long-focus telescopes, which provided greater magnification and resolution than earlier models. These advancements allowed Cassini and his contemporaries to make more detailed observations of celestial objects, paving the way for future discoveries.

Throughout his career, Cassini maintained a deep interest in the practical applications of astronomy. He was involved in several important projects related to geodesy, the science of measuring and understanding the Earth's geometric shape, orientation in space, and gravitational field. One of his notable achievements in this area was his work on the measurement of the meridian arc between Paris and the French city of Amiens. This project, completed in 1718, was a significant milestone in the development of more accurate maps and the understanding of the Earth's shape.

Cassini's contributions to the field of astronomy were widely recognized and celebrated during his lifetime. He received numerous honors and accolades, including membership in the prestigious French Academy of Sciences. His work laid the foundation for future generations of astronomers and had a lasting impact on the development of the field.

Despite his many achievements, Cassini's career was not without controversy. His opposition to the heliocentric model of the solar system, which placed the Sun at the center rather than the Earth, put him at odds with some of his contemporaries. Cassini initially supported the Tychonic system, a geocentric model proposed by the Danish astronomer Tycho Brahe, which combined elements of both the Ptolemaic and Copernican systems. Over time, however, the evidence in favor of the heliocentric model became overwhelming, and Cassini's views evolved accordingly.

Cassini's legacy continued through his descendants, who made significant contributions to the field of astronomy. His son, Jacques Cassini, succeeded him as the director of the Paris Observatory and continued his work on planetary observations and geodesy. Jacques' son, César-François Cassini de Thury, and grandson, Jean-Dominique Cassini, also became prominent astronomers, carrying on the family tradition of scientific inquiry and discovery.

The Cassini name remains synonymous with astronomical excellence. In 1997, NASA launched the Cassini spacecraft, a joint mission with the European Space Agency (ESA) and the Italian Space Agency (ASI) to explore Saturn and its moons. The Cassini spacecraft, named in honor of Giovanni Cassini, provided unprecedented data and images of the Saturnian system, revolutionizing our understanding of the planet and its complex ring system. The mission's numerous discoveries, including detailed studies of Saturn's atmosphere, rings, and moons, are a fitting tribute to Cassini's pioneering work and enduring legacy.

Giovanni Cassini's contributions to the field of astronomy are vast and multifaceted. His discoveries of Saturn's moons and the Cassini Division in the planet's rings have had a lasting impact on our understanding of the solar system. His work in astrometry, lunar mapping, and geodesy laid the groundwork for future advancements in these fields. Cassini's commitment to scientific inquiry, innovation, and precision has inspired generations of astronomers and continues to influence the field of astronomy today.

Cassini's life and career exemplify the spirit of discovery and the pursuit of knowledge. His dedication to advancing our understanding of the universe, combined with his innovative approach to observation and measurement, has left an indelible mark on the field of astronomy. Giovanni Cassini's legacy is one of exploration, innovation, and a deep curiosity about the cosmos, and his contributions will continue to be celebrated and studied for generations to come.

Chapter 5: Isaac Newton

Isaac Newton, born on January 4, 1643, in Woolsthorpe, Lincolnshire, England, stands as one of the most influential figures in the history of science. His contributions to mathematics, physics, astronomy, and natural philosophy laid the foundations for classical mechanics and profoundly influenced the scientific revolution. Newton's life and work exemplify the spirit of inquiry and the pursuit of knowledge that characterized this transformative period.

Newton's early life was marked by hardship and solitude. His father, also named Isaac Newton, died three months before his birth, and his mother, Hannah Ayscough Newton, remarried and left young Isaac in the care of his maternal grandmother. This separation had a profound impact on Newton, fostering a sense of independence and self-reliance. Despite a turbulent childhood, Newton showed an early aptitude for mechanics and an intense curiosity about the natural world.

In 1661, Newton entered Trinity College, Cambridge, where he was introduced to the works of contemporary philosophers and scientists, including René Descartes, Galileo Galilei, and Johannes Kepler. Under the tutelage of Isaac Barrow, a leading mathematician, Newton quickly distinguished himself. During his time at Cambridge, Newton began his work on calculus, optics, and the laws of motion and gravitation that would later define his legacy.

Newton's most productive period began in 1665, when the outbreak of the Great Plague forced Cambridge to close, and he returned to Woolsthorpe. During this period of isolation, Newton made remarkable strides in several areas of science and mathematics. This time is often referred to as his "annus mirabilis" or "year of wonders." It was during this period that Newton developed the foundations of calculus, an essential mathematical tool that allows for the precise calculation of rates of change and the summation of infinite series. Independently, German mathematician Gottfried Wilhelm Leibniz developed a similar system, leading to a bitter dispute over the priority of the invention that lasted for many years.

In addition to his work on calculus, Newton made significant advances in optics. He conducted a series of experiments with prisms, demonstrating that white light is composed of a spectrum of colors, each with a different refrangibility. This work challenged the prevailing theory that colors were modifications of white light and laid the groundwork for the field of spectroscopy. Newton published his findings in "Opticks," a comprehensive work that detailed his experiments and theories on the nature of light and color.

Newton's most famous contributions to science are encapsulated in his magnum opus, "Philosophiæ Naturalis Principia Mathematica" (Mathematical Principles of Natural Philosophy), commonly known as the "Principia." Published in 1687, the "Principia" introduced the three laws of motion and the universal law of gravitation, which together provided a unified framework for understanding the physical universe.

Newton's first law of motion, the law of inertia, states that an object at rest will remain at rest, and an object in motion will remain in motion at a constant velocity unless acted upon by an external force. This principle challenged the Aristotelian view that a force is required to maintain motion and established the concept of inertia as a fundamental property of matter.

The second law of motion, the law of acceleration, quantifies the relationship between force, mass, and acceleration. It states that the force acting on an object is equal to the mass of the object multiplied by its acceleration ($F = ma$). This law provided a mathematical description of how forces influence the motion of objects and allowed for precise predictions of their behavior under various conditions.

Newton's third law of motion, the law of action and reaction, asserts that for every action, there is an equal and opposite reaction. This principle explains the interactions between objects and is foundational to the study of dynamics and the behavior of systems in motion.

In addition to his laws of motion, Newton formulated the universal law of gravitation. He proposed that every particle of matter in the universe attracts every other particle with a force that is directly proportional to the product of their masses and inversely proportional to the square of the distance between their centers ($F = G(m1m2)/r^2$). This law explained the motion of celestial bodies and provided a framework for understanding phenomena such as planetary orbits, tides, and the behavior of objects on Earth.

Newton's gravitational theory was revolutionary, providing a unified explanation for both terrestrial and celestial phenomena. It resolved long-standing questions about the nature of planetary motion, building on the work of Johannes Kepler and his laws of planetary motion. Newton's synthesis of mathematical principles and empirical observations transformed natural philosophy into a rigorous, quantitative science.

Beyond his contributions to physics and mathematics, Newton also made significant advances in astronomy. He invented the reflecting telescope, which used a curved mirror to focus light and produce clearer images than the refracting telescopes of his time. This innovation allowed for more precise observations of celestial bodies and contributed to the advancement of astronomical research.

Newton's later years were marked by his involvement in various administrative and political roles. In 1696, he was appointed Warden of the Royal Mint, where he played a crucial role in reforming the English currency and combating counterfeiting. His rigorous approach to the task and his commitment to maintaining the integrity of the currency earned him considerable respect and recognition.

In 1703, Newton was elected President of the Royal Society, a position he held until his death. In this role, he promoted scientific inquiry and the dissemination of knowledge, fostering a spirit of collaboration and innovation among the scientific community. His tenure as president solidified his status as a leading figure in the scientific world and a champion of the empirical method.

Despite his monumental achievements, Newton's life was not without controversy and personal challenges. His dispute with Leibniz over the invention of calculus was particularly acrimonious, leading to a prolonged and bitter rivalry that divided the mathematical community. Additionally, Newton's later years were marked by periods of ill health and a preoccupation with alchemical and theological studies, which some contemporaries viewed as eccentric.

Newton's influence extended beyond his scientific contributions. His work profoundly shaped the Enlightenment, a period characterized by the emphasis on reason, empirical evidence, and the pursuit of knowledge. Newton's principles of natural philosophy provided a framework for understanding the natural world that was both rational and predictive, inspiring future generations of scientists and thinkers.

Isaac Newton died on March 31, 1727, and was buried in Westminster Abbey, an honor befitting his monumental contributions to science and humanity. His legacy endures through the continued application of his principles in physics, mathematics, astronomy, and engineering. Newton's work laid the foundation for classical mechanics, influencing the development of modern physics and shaping our understanding of the universe.

Newton's impact on science and human knowledge cannot be overstated. His meticulous approach to experimentation, his ability to synthesize complex ideas, and his unwavering commitment to uncovering the truths of the natural world set a standard for scientific inquiry that endures to this day. Newton's laws of motion and universal gravitation remain fundamental to our understanding of the physical universe, and his innovations in mathematics and optics continue to influence contemporary research and technology.

Isaac Newton's life and work exemplify the transformative power of scientific inquiry and the relentless pursuit of knowledge. His contributions to science revolutionized our understanding of the natural world and laid the groundwork for countless advancements in technology and industry. Newton's legacy is a testament to the enduring importance of curiosity, rigor, and the quest for understanding that drives the scientific endeavor.

Chapter 6: Konstantin Tsiolkovsky

Konstantin Eduardovich Tsiolkovsky, born on September 17, 1857, in the village of Izhevskoye, in the Russian Empire, is widely regarded as one of the founding fathers of astronautics and rocket science. His visionary ideas and theoretical work laid the groundwork for space exploration and greatly influenced subsequent developments in the field. Tsiolkovsky's life and work exemplify the power of imagination and scientific inquiry, demonstrating how a deep understanding of physics and engineering can transcend the limitations of the time to envision the future.

Tsiolkovsky was born into a family of modest means. His father, Eduard Ignatyevich Tsiolkovsky, was a Polish-born forest ranger, and his mother, Maria Ivanovna Yumasheva, was of Russian descent. At the age of ten, Tsiolkovsky contracted scarlet fever, which resulted in a significant hearing loss. This disability profoundly affected his education, as he found it difficult to follow lessons in a conventional classroom setting. Consequently, he was largely self-taught, developing a voracious appetite for reading and a keen interest in science and mathematics.

As a young man, Tsiolkovsky moved to Moscow, where he immersed himself in the city's libraries, studying a wide range of scientific literature. He was particularly inspired by the works of Jules Verne, whose science fiction novels fueled his imagination and aspirations for space travel. Despite his lack of formal education, Tsiolkovsky's relentless self-study and natural aptitude for mathematics and physics enabled him to develop a solid foundation in these disciplines.

In 1879, Tsiolkovsky passed the necessary examinations to become a teacher and began working as a schoolteacher in Borovsk, a small town near Moscow. He continued to teach for much of his life, eventually moving to Kaluga, where he spent the remainder of his career. Teaching provided him with a stable income and allowed him to pursue his scientific interests in his spare time.

Tsiolkovsky's early scientific work focused on aeronautics. He conducted experiments with wind tunnels and developed designs for various aircraft, including dirigibles and airplanes. His interest in aeronautics naturally extended to the concept of space travel, and he began to explore the theoretical foundations of rocketry. In 1898, Tsiolkovsky published his seminal work, "Exploration of Outer Space by Means of Rocket Devices." In this paper, he introduced the idea of using rockets for space travel and formulated the basic principles of rocket propulsion.

One of Tsiolkovsky's most significant contributions to astronautics was the development of the rocket equation, also known as the Tsiolkovsky equation. This mathematical formula describes the motion of a rocket as it expels mass in the form of exhaust gases. The equation relates the change in velocity of the rocket to the velocity of the exhaust gases and the initial and final mass of the rocket. The Tsiolkovsky equation is fundamental to rocketry and space travel, as it provides a quantitative understanding of how rockets achieve propulsion and reach high velocities.

Tsiolkovsky's work also emphasized the importance of multi-stage rockets. He recognized that a single-stage rocket would be insufficient to achieve the velocities necessary for space travel due to the limitations imposed by the mass of the fuel and the rocket structure. Instead, he proposed the use of multiple stages, each of which would be jettisoned after its fuel was expended, thereby reducing the mass of the rocket and allowing for greater efficiency. This concept of staging is a cornerstone of modern rocketry and is employed in virtually all space launch vehicles.

In addition to his theoretical work on rocketry, Tsiolkovsky made significant contributions to the understanding of spaceflight mechanics. He explored the challenges of human space travel, including the effects of microgravity and the need for life support systems. He proposed designs for spacecraft that included features such as sealed cabins, oxygen supplies, and temperature control mechanisms. Tsiolkovsky's visionary ideas about space habitats and colonization anticipated many of the concepts that would later become central to space exploration.

Tsiolkovsky's vision extended beyond the technical aspects of rocketry and space travel. He was deeply interested in the philosophical and ethical implications of humanity's expansion into space. He viewed space exploration as a means of ensuring the survival and progress of humanity, which he saw as an inevitable step in the evolution of human civilization. Tsiolkovsky believed that the colonization of other planets would provide new opportunities for social and technological development, as well as a safeguard against potential existential threats on Earth.

Despite his groundbreaking work, Tsiolkovsky received little recognition or support during his lifetime. He worked in relative isolation, with limited access to resources and collaborators. His publications were often self-funded, and he faced significant skepticism from the scientific community and the general public. Nevertheless, Tsiolkovsky remained dedicated to his research and continued to develop his ideas and theories.

In the 1920s and 1930s, as the field of rocketry began to gain more attention, Tsiolkovsky's contributions started to receive greater recognition. Soviet authorities, recognizing the potential strategic importance of rocketry, began to support and promote his work. Tsiolkovsky was honored with several awards and titles, and his writings were published and disseminated more widely. He became a symbol of Soviet scientific achievement and an inspiration to a new generation of engineers and scientists.

Tsiolkovsky's influence on the development of rocketry and space exploration cannot be overstated. His theoretical work laid the groundwork for many of the advances that followed, and his visionary ideas inspired countless engineers and scientists. Among those influenced by Tsiolkovsky was Sergei Korolev, the chief architect of the Soviet space program, who played a pivotal role in launching the first artificial satellite, Sputnik, and sending the first human, Yuri Gagarin, into space.

Tsiolkovsky's legacy extends beyond his contributions to rocketry and space travel. His philosophical and ethical reflections on space exploration continue to resonate with those who advocate for humanity's expansion into the cosmos. Tsiolkovsky's belief in the potential for space colonization to drive social and technological progress has inspired numerous initiatives aimed at exploring and settling other planets.

In recognition of his contributions, Tsiolkovsky has been honored in various ways. Numerous streets, schools, and institutions in Russia bear his name, and he is commemorated by statues and monuments. The Tsiolkovsky State Museum of the History of Cosmonautics in Kaluga, founded in 1967, serves as a testament to his enduring legacy and a repository of his life's work.

Tsiolkovsky's ideas have also influenced the broader field of science fiction, inspiring writers and filmmakers to explore the possibilities of space travel and human expansion into the cosmos. His visionary concepts, such as space stations, interplanetary travel, and space colonization, have become staples of the genre and continue to capture the imagination of audiences worldwide.

Chapter 7: Georges Lemaître

Georges Henri Joseph Édouard Lemaître, born on July 17, 1894, in Charleroi, Belgium, was a Belgian Catholic priest, astronomer, and professor of physics who is best known for formulating the modern Big Bang theory of the origin of the universe. His groundbreaking work on the expansion of the universe laid the foundation for modern cosmology and significantly advanced our understanding of the cosmos. Lemaître's life and contributions exemplify the harmonious relationship between faith and science, demonstrating how deep religious beliefs can coexist with rigorous scientific inquiry.

Lemaître was born into a devout Catholic family. His father, Joseph Lemaître, was an industrialist, and his mother, Marguerite Lannoy, was deeply religious. From an early age, Lemaître displayed a keen interest in science and mathematics. He was educated at a Jesuit secondary school, where he excelled in his studies and developed a strong foundation in both scientific and religious thought.

In 1911, Lemaître enrolled at the Catholic University of Leuven (Université catholique de Louvain), where he studied civil engineering. However, his education was interrupted by World War I. Lemaître served as an artillery officer in the Belgian army and was awarded the Belgian War Cross for his bravery. The war had a profound impact on him, deepening his religious convictions and his commitment to understanding the universe.

After the war, Lemaître returned to Leuven to continue his studies, but he shifted his focus to mathematics and physics. He earned a doctorate in 1920 with a thesis on the approximation of functions of several real variables. Simultaneously, Lemaître pursued his theological studies and was ordained a Catholic priest in 1923. This unique combination of scientific and theological education set the stage for his future contributions to cosmology.

Lemaître's interest in cosmology was sparked by the work of Albert Einstein and the development of general relativity. Einstein's equations described how gravity affects the fabric of space-time, providing a new framework for understanding the universe. Lemaître recognized the potential of general relativity to describe the dynamics of the cosmos and began to explore its implications.

In 1923, Lemaître received a fellowship to study abroad, which allowed him to spend time at the University of Cambridge, where he studied under the eminent astronomer Arthur Eddington. Eddington's work on the relationship between the theory of relativity and stellar structure greatly influenced Lemaître. He then spent a year at Harvard College Observatory and the Massachusetts Institute of Technology, where he continued to expand his knowledge of astrophysics and cosmology.

In 1927, Lemaître published a groundbreaking paper in the Annales de la Société Scientifique de Bruxelles, titled "Un Univers homogène de masse constante et de rayon croissant rendant compte de la vitesse radiale des nébuleuses extra-galactiques" (A Homogeneous Universe of Constant Mass and Increasing Radius Accounting for the Radial Velocity of Extra-Galactic Nebulae). In this paper, Lemaître proposed that the universe is expanding, a revolutionary idea that challenged the prevailing static universe model. He derived what is now known as Hubble's Law, which relates the recessional velocity of galaxies to their distance from Earth, indicating that galaxies are moving away from each other as the universe expands.

Lemaître's work initially received little attention outside of Belgium, partly due to the language barrier and partly because the scientific community was still largely committed to the idea of a static universe. However, his ideas gained significant support in the early 1930s, particularly after Edwin Hubble published his observations of the redshifts of galaxies, which provided empirical evidence for Lemaître's theoretical predictions. Hubble's observations confirmed that distant galaxies were indeed moving away from us, lending strong support to the expanding universe model.

Lemaître went further in his exploration of the origins of the universe. In 1931, he proposed what he called the "hypothesis of the primeval atom," which suggested that the universe began as a single, extremely hot and dense point, which he referred to as the "cosmic egg" that exploded to give birth to the expanding universe we observe today. This idea was a precursor to what is now known as the Big Bang theory. Lemaître's primeval atom hypothesis was based on the principles of thermodynamics and quantum mechanics, and it provided a plausible explanation for the observed expansion of the universe.

Lemaître's hypothesis initially faced resistance from several quarters, including from Einstein himself. When Lemaître first presented his expanding universe theory to Einstein in 1931, Einstein reportedly dismissed it, saying, "Your calculations are correct, but your physics is abominable." However, Einstein later revised his stance and praised Lemaître's work after further evidence supporting the expanding universe model became apparent. Einstein is said to have acknowledged the beauty of Lemaître's theory, reportedly saying, "This is the most beautiful and satisfactory explanation of creation to which I have ever listened."

Lemaître continued to refine his cosmological theories throughout his career. He explored the implications of an expanding universe for the formation of galaxies and the large-scale structure of the cosmos. He also investigated the role of cosmic radiation and the potential for relic radiation from the early universe, ideas that would later be confirmed with the discovery of the cosmic microwave background radiation in 1965 by Arno Penzias and Robert Wilson. This discovery provided strong evidence for the Big Bang theory and solidified Lemaître's place in the history of cosmology.

Despite his monumental contributions to science, Lemaître remained deeply committed to his religious faith. He saw no conflict between his scientific work and his religious beliefs. In fact, he believed that his scientific discoveries revealed the grandeur of God's creation. Lemaître was a member of the Pontifical Academy of Sciences and served as its president from 1960 until his death in 1966. He maintained a humble and unassuming demeanor, always emphasizing the provisional nature of scientific knowledge and the importance of faith.

Lemaître's contributions to cosmology were recognized by numerous awards and honors during his lifetime. He was elected a member of the Royal Academy of Science, Letters, and Fine Arts of Belgium, and he received honorary degrees from several universities. In 1934, he was awarded the Francqui Prize, the highest scientific honor in Belgium. In 1953, Lemaître received the inaugural Eddington Medal from the Royal Astronomical Society, named after his mentor, Arthur Eddington.

Lemaître's influence extended beyond his scientific achievements. His work inspired future generations of cosmologists and astrophysicists, including notable figures such as George Gamow, who further developed the Big Bang theory and the concept of nucleosynthesis in the early universe. Lemaître's pioneering ideas also paved the way for the development of modern observational cosmology, including the study of large-scale structure, dark matter, and dark energy.

Lemaître's legacy is also reflected in the numerous institutions and awards named in his honor. The Georges Lemaître Centre for Earth and Climate Research at the Catholic University of Leuven, the Georges Lemaître Prize awarded by the Université catholique de Louvain, and the Lemaître Coordinates used in cosmological models all bear his name. These honors testify to the enduring impact of his work on the field of cosmology and the broader scientific community.

Lemaître's life and work continue to inspire scholars and thinkers across disciplines. His ability to bridge the gap between science and religion, and his unwavering commitment to the pursuit of knowledge, serve as a powerful example of the potential for human inquiry to transcend traditional boundaries. Lemaître's story is a testament to the power of curiosity, perseverance, and faith in the quest for understanding the fundamental nature of the universe.

Chapter 8: Robert H. Goddard

Robert Hutchings Goddard, born on October 5, 1882, in Worcester, Massachusetts, is often hailed as the father of modern rocketry. His pioneering work in the early 20th century laid the foundation for the development of the rockets that would eventually take humans to the Moon and explore the far reaches of our solar system. Goddard's relentless pursuit of his vision, despite considerable skepticism and numerous challenges, has left an indelible mark on the field of astronautics and space exploration.

Goddard's early life was marked by a deep fascination with science and a voracious appetite for learning. As a young boy, he was captivated by stories of space travel and exploration, particularly H.G. Wells' science fiction novel "The War of the Worlds." This interest in space and science fiction sparked his imagination and led him to dream about the possibility of traveling beyond the Earth. His curiosity was further fueled by his father's encouragement to experiment with various scientific concepts, including electricity and chemistry.

In 1899, at the age of 17, Goddard experienced what he later described as his "Anniversary Day," a defining moment in his life. While climbing a cherry tree in his backyard, he envisioned a device that could reach Mars. This epiphany profoundly influenced his future work and solidified his determination to develop the technology needed for space travel. He marked this day, October 19, as the anniversary of his life's purpose and referred to it throughout his career.

Goddard's formal education began at Worcester Polytechnic Institute, where he majored in physics and graduated in 1908. His academic achievements earned him a scholarship to study at Clark University, where he completed his Ph.D. in physics in 1911. At Clark, he began to formulate the theoretical foundations for his work in rocketry. His doctoral thesis, "On a Method of Reaching Extreme Altitudes," was a seminal work that laid out the principles of rocket propulsion and the use of liquid fuels to achieve greater efficiency and higher altitudes.

In his thesis, Goddard demonstrated an advanced understanding of the physics of rocketry. He proposed the use of liquid hydrogen and liquid oxygen as propellants, which would provide a much higher specific impulse than the solid propellants used in earlier rockets. He also calculated the theoretical altitude that could be reached by a rocket using these propellants and explored the challenges of overcoming atmospheric drag and gravitational forces. His work was pioneering in its thoroughness and its vision of future space travel.

Despite the groundbreaking nature of his work, Goddard faced significant skepticism from the scientific community and the public. Many contemporaries doubted the feasibility of rocketry for space travel, and his ideas were often met with ridicule. Undeterred, Goddard continued his research, securing a position as a physics professor at Clark University, which provided him with a stable income and access to laboratory facilities.

In 1914, Goddard received two critical patents: one for a multi-stage rocket and another for a liquid-fuel rocket. These patents were among the first to detail the use of liquid propellants and staging in rocketry, concepts that would later become fundamental to modern spaceflight. His work attracted the attention of the Smithsonian Institution, which provided him with a grant in 1917 to further his research. With this funding, Goddard was able to build and test small-scale rockets, refining his designs and experimenting with different fuels and materials.

World War I temporarily shifted Goddard's focus to military applications of rocketry. He developed a shoulder-fired rocket-propelled grenade, known as the "bazooka," which saw limited use in the war. Although his work during this period did not directly contribute to space exploration, it provided valuable experience in rocket design and testing.

After the war, Goddard returned to his pursuit of high-altitude rocketry. In 1920, he published a report titled "A Method of Reaching Extreme Altitudes," in which he outlined his vision for space travel and the potential for rockets to reach the Moon. This report, funded by the Smithsonian Institution, included detailed calculations and experimental data supporting his theories. However, it also attracted public ridicule, particularly for his suggestion that rockets could operate in the vacuum of space. The New York Times famously mocked him, questioning his understanding of basic physics.

Despite the public skepticism, Goddard persevered. In 1926, he achieved a major milestone by successfully launching the world's first liquid-fueled rocket. On March 16, 1926, in Auburn, Massachusetts, Goddard's rocket, which he named "Nell," ascended to an altitude of 41 feet and traveled a distance of 184 feet. Although modest by today's standards, this flight marked a significant breakthrough in rocketry. It demonstrated the feasibility of liquid-fueled propulsion and provided proof of concept for Goddard's theories.

Following this success, Goddard continued to refine his rocket designs and conduct more advanced experiments. He moved his operations to Roswell, New Mexico, in 1930, seeking a more secluded and suitable environment for his tests. With funding from the Guggenheim Foundation and support from wealthy benefactors like Charles Lindbergh, Goddard established a research facility in the remote desert. There, he and his team conducted numerous test flights, achieving higher altitudes and greater distances.

One of Goddard's significant contributions during this period was the development of gyroscopic stabilization and guidance systems for rockets. He recognized that accurate control of a rocket's flight path was essential for achieving high altitudes and reaching specific targets. By incorporating gyroscopes and other control mechanisms into his designs, Goddard was able to improve the stability and precision of his rockets, laying the groundwork for modern guidance systems.

Goddard's work in Roswell also included experiments with different fuel mixtures, nozzle designs, and cooling systems. He developed regenerative cooling, a method in which the rocket's fuel is circulated around the combustion chamber to absorb heat and prevent overheating. This innovation allowed for more efficient and reliable operation of liquid-fueled rockets, addressing one of the critical challenges in rocket design.

Despite his many achievements, Goddard continued to work in relative obscurity. His research was largely self-funded, and he received little recognition or support from the government or the broader scientific community. However, his contributions did not go unnoticed by everyone. German rocket scientists, including Wernher von Braun, closely followed Goddard's work and incorporated many of his innovations into their own rocket programs. After World War II, von Braun acknowledged Goddard's influence, stating that Goddard's patents and publications had significantly advanced their understanding of rocketry.

Goddard's health began to decline in the late 1930s, and he was diagnosed with throat cancer in 1944. Despite his illness, he continued to work on his research until his death on August 10, 1945. Although he did not live to see the full realization of his vision, his pioneering work laid the foundation for the space age.

In the years following his death, Goddard's contributions to rocketry and space exploration were increasingly recognized and celebrated. In 1951, the U.S. government awarded his widow, Esther Goddard, and the Guggenheim Foundation $1 million in compensation for the use of his patents during World War II. His work was instrumental in the development of the V-2 rocket by the Germans, which in turn influenced the design of the rockets used in the U.S. space program.

The legacy of Robert H. Goddard is enshrined in numerous honors and memorials. NASA's Goddard Space Flight Center in Greenbelt,

Maryland, named in his honor, serves as a testament to his enduring impact on space exploration. The American Institute of Aeronautics and Astronautics (AIAA) awards the Robert H. Goddard Memorial Trophy annually to recognize outstanding achievements in rocketry and astronautics.

Goddard's contributions to science and engineering extend beyond his technical innovations. His work exemplifies the spirit of perseverance and dedication in the face of adversity. Despite significant skepticism and numerous setbacks, Goddard remained committed to his vision of space travel, continually pushing the boundaries of what was considered possible. His story serves as an inspiration to scientists and engineers, demonstrating the importance of pursuing one's dreams with passion and tenacity.

In addition to his technical achievements, Goddard's work had profound philosophical implications. He envisioned a future in which humanity would explore and colonize other planets, expanding the reach of human civilization beyond the confines of Earth. His vision of space travel as a means of ensuring the survival and progress of humanity continues to resonate with those who advocate for space exploration.

Robert H. Goddard's life and work represent a remarkable blend of scientific ingenuity, visionary thinking, and unwavering determination. His pioneering contributions to rocketry laid the foundation for the space age, enabling humanity to explore the cosmos and achieve remarkable feats in space exploration. Despite the challenges he faced, Goddard's legacy endures, inspiring future generations to reach for the stars and push the boundaries of human knowledge and capability.

Chapter 9: Wernher von Braun

Wernher von Braun, born on March 23, 1912, in Wirsitz, Prussia (now Wyrzysk, Poland), was one of the most influential figures in the history of rocketry and space exploration. His work spanned several decades and had a profound impact on both the German V-2 rocket program during World War II and the American space program during the Cold War. Von Braun's career is a testament to his extraordinary vision and technical prowess, though it is also marked by controversy due to his involvement with the Nazi regime.

Von Braun was born into an aristocratic family, the second of three sons of Magnus von Braun, a former East Prussian Minister of Agriculture, and his wife, Emmy von Quistorp. From a young age, von Braun displayed a keen interest in science and engineering, particularly in rocketry and space travel. His early fascination with space was fueled by the science fiction works of Jules Verne and H.G. Wells, as well as the technical writings of Hermann Oberth, one of the founding fathers of rocketry.

In 1930, von Braun enrolled at the Technical University of Berlin, where he studied mechanical engineering and worked with Oberth on liquid-fueled rocket experiments. This collaboration marked the beginning of his professional career in rocketry. Von Braun earned his doctorate in physics in 1934 with a thesis on liquid-fueled rocket engines, laying the groundwork for his future innovations.

Von Braun's early work in rocketry caught the attention of the German military, which saw the potential of rockets as weapons. In 1932, he joined the German Army Ordnance Office, where he led a team of engineers and scientists in developing rocket technology for military purposes. This collaboration resulted in the creation of the Aggregate series of rockets, culminating in the A-4, better known as the V-2 rocket.

The V-2 rocket, or Vergeltungswaffe 2 (Vengeance Weapon 2), was the world's first long-range guided ballistic missile. It was powered by a liquid-fueled rocket engine and capable of delivering a one-ton warhead over a distance of 200 miles. The V-2 represented a significant technological achievement and marked a major milestone in the development of rocketry. However, its use as a weapon of terror against civilian populations in Europe, particularly in London and Antwerp, remains a dark chapter in von Braun's legacy.

The V-2 program was headquartered at the Peenemünde Army Research Center on the Baltic coast, where von Braun and his team conducted extensive research and development. Despite the technical success of the V-2, the program was marred by its association with the Nazi regime and the use of forced labor from concentration camps to build the rockets. Thousands of prisoners died under horrific conditions in the Mittelwerk underground factory, where the rockets were produced. Von Braun's knowledge of and involvement in these atrocities is a subject of historical debate and ethical scrutiny.

As World War II drew to a close, von Braun and his team sought to surrender to the American forces, fearing capture by the Soviets. In May 1945, von Braun and around 500 of his colleagues, along with their families, surrendered to the U.S. Army in Bavaria. This strategic move allowed them to avoid prosecution for war crimes and paved the way for their future contributions to the American space program.

Under Operation Paperclip, a secret program to recruit German scientists, von Braun and his team were brought to the United States. Initially, they were taken to Fort Bliss, Texas, where they worked on rocketry projects for the U.S. Army. This transition marked the beginning of a new chapter in von Braun's career, as he shifted his focus from military applications to space exploration.

In 1950, von Braun and his team moved to Huntsville, Alabama, where they continued their work on rocket development at the Redstone Arsenal. One of their notable achievements was the development of the Redstone rocket, a modified version of the V-2 that became the first large American ballistic missile. The Redstone rocket also played a crucial role in the early stages of the U.S. space program, serving as the launch vehicle for the first American satellite, Explorer 1, in 1958.

Von Braun's vision of space exploration extended beyond Earth orbit. He was a passionate advocate for human spaceflight and the colonization of other planets. In the early 1950s, he began collaborating with Walt Disney on a series of television programs that popularized the concept of space travel and inspired public interest in the space race. These programs, which included titles such as "Man in Space" and "Mars and Beyond," showcased von Braun's designs for space stations, lunar bases, and manned missions to Mars.

The culmination of von Braun's work came with his leadership of NASA's Marshall Space Flight Center in Huntsville. Appointed as its first director in 1960, he played a pivotal role in the development of the Saturn V rocket, the launch vehicle that would ultimately take astronauts to the Moon. The Saturn V was a colossal achievement in engineering, standing 363 feet tall and capable of delivering over 6 million pounds of thrust. It remains the most powerful rocket ever built and was instrumental in achieving the goal set by President John F. Kennedy in 1961 to land a man on the Moon before the end of the decade.

The Apollo program, powered by the Saturn V, marked a turning point in human history. On July 20, 1969, Apollo 11 successfully landed on the Moon, with astronauts Neil Armstrong and Edwin "Buzz" Aldrin becoming the first humans to set foot on another celestial body. Von Braun's contributions to this monumental achievement cemented his legacy as one of the foremost pioneers of space exploration.

Despite his technical achievements, von Braun's past continued to haunt him. His involvement with the Nazi regime and the use of forced labor in the V-2 program remained controversial. In the postwar years, von Braun publicly distanced himself from his wartime activities, emphasizing his commitment to space exploration and peaceful uses of rocket technology. However, the ethical implications of his work during the war remain a subject of ongoing debate.

In the later years of his career, von Braun continued to advocate for space exploration. He envisioned a future in which humanity would establish colonies on the Moon and Mars, and he proposed ambitious plans for interplanetary missions. However, his vision was tempered by the realities of budget constraints and shifting political priorities. In 1970, von Braun was transferred to NASA headquarters in Washington, D.C., where he served as Deputy Associate Administrator for Planning. He retired from NASA in 1972 and subsequently joined Fairchild Industries, where he continued to promote space exploration.

Von Braun's health began to decline in the mid-1970s, and he was diagnosed with kidney cancer in 1976. Despite his illness, he remained active in advocating for space exploration until his death on June 16, 1977. His passing marked the end of an era in the history of rocketry and space exploration.

Wernher von Braun's legacy is multifaceted and complex. On one hand, he is celebrated as a visionary scientist and engineer whose work enabled humanity to reach the Moon and explore the cosmos. His contributions to rocketry and spaceflight have had a lasting impact on science, technology, and our understanding of the universe. On the other hand, his association with the Nazi regime and the ethical questions surrounding his wartime activities continue to cast a shadow over his achievements.

In recognition of his contributions, von Braun has been honored with numerous awards and accolades. He received the National Medal of Science in 1975, and his name is commemorated in various institutions and landmarks, including the Wernher von Braun Center for Science & Innovation in Huntsville, Alabama. His story serves as a powerful reminder of the complex interplay between science, ethics, and history.

Von Braun's influence extends beyond his technical achievements. He played a crucial role in shaping the public perception of space exploration and inspiring future generations of scientists, engineers, and explorers. His collaboration with Walt Disney and his advocacy for space travel helped to ignite the imagination of millions and foster a sense of wonder and possibility about the universe.

Chapter 10: Sergei Korolev

Sergei Pavlovich Korolev was a pivotal figure in the history of space exploration, recognized as the chief architect behind the Soviet space program. Born on January 12, 1907, in Zhytomyr, Ukraine, then part of the Russian Empire, Korolev's early life was marked by his parents' separation. He was raised by his mother and maternal grandparents. From a young age, Korolev exhibited a keen interest in aviation and engineering, which steered him towards a career that would eventually impact the course of human history.

Korolev pursued his education in the field of aeronautical engineering, studying at the Kyiv Polytechnic Institute and later at the Bauman Moscow State Technical University. His talents were recognized early on, and he became involved in the Soviet Union's nascent rocketry programs in the 1930s. Working alongside other pioneering engineers like Friedrich Zander and Mikhail Tikhonravov, Korolev co-founded the Group for the Study of Reactive Motion (GIRD), which was instrumental in advancing rocket technology in the USSR.

In 1933, Korolev and his colleagues successfully launched the GIRD-X, one of the Soviet Union's first liquid-fueled rockets. This achievement marked a significant milestone and laid the groundwork for future developments. However, Korolev's career took a dramatic turn during the Great Purge under Joseph Stalin's regime. In 1938, he was arrested on charges of espionage and sabotage, likely due to the paranoid atmosphere and widespread purges of the time. Korolev endured harsh conditions in the Gulag, spending time in the Kolyma gold mines in Siberia, one of the most brutal labor camps.

Korolev's fortunes changed during World War II, when the Soviet Union recognized the strategic importance of rocket technology. He was released from the Gulag in 1944, albeit without a full pardon, and brought back to work on military rocket development. His expertise proved invaluable in the development of the R-7 Semyorka, the world's first intercontinental ballistic missile (ICBM). This rocket would later serve as the basis for launching the first artificial satellite, Sputnik, into space.

As the chief designer of the Soviet space program, Korolev was responsible for numerous groundbreaking achievements. Under his leadership, the Soviet Union launched Sputnik 1 on October 4, 1957, marking the beginning of the space age and the start of the space race with the United States. This success was followed by a series of remarkable feats, including the launch of the first human, Yuri Gagarin, into space on April 12, 1961. Gagarin's successful orbit of the Earth aboard Vostok 1 was a triumph for the Soviet space program and solidified Korolev's legacy as a pioneering figure in space exploration.

Korolev's vision extended beyond manned spaceflight. He played a crucial role in the development of the Luna program, which achieved the first successful landing of an unmanned spacecraft on the Moon with Luna 2 in 1959. The Luna missions also included the first photographs of the far side of the Moon, captured by Luna 3, and the first successful return of lunar soil samples by Luna 16.

Despite these successes, Korolev faced numerous challenges and setbacks. The Soviet space program was marked by intense secrecy, internal political struggles, and competition with the United States' NASA. Korolev's work was often hampered by limited resources and the need to balance military and scientific objectives. Nevertheless, his ability to inspire and lead a team of engineers and scientists was instrumental in overcoming these obstacles.

Korolev's contributions extended beyond technological advancements; he also fostered a spirit of exploration and international competition that pushed the boundaries of what humanity could achieve. His emphasis on innovation, engineering excellence, and the pursuit of ambitious goals inspired future generations of scientists and engineers.

Tragically, Korolev's life was cut short at the age of 59. He died on January 14, 1966, following complications from surgery. His death was a significant loss to the Soviet space program, which struggled to maintain its momentum in the years that followed. Korolev was buried with honors in the Kremlin Wall Necropolis, a testament to his status as a national hero.

In the years since his death, Korolev's contributions have been more widely recognized and celebrated. The secrecy surrounding his work during his lifetime meant that much of his role in the Soviet space program only became known posthumously. Today, he is remembered as one of the foremost figures in the history of space exploration, often compared to Wernher von Braun, his counterpart in the United States.

Korolev's legacy endures in the continued exploration of space. The launch vehicles and technologies he developed laid the foundation for subsequent space missions, both manned and unmanned. His vision of space exploration as a unifying and inspiring endeavor continues to resonate, reminding us of the extraordinary achievements that are possible when human ingenuity and determination are applied to the pursuit of knowledge and discovery.

Chapter 11: Edwin Hubble

Edwin Powell Hubble was an American astronomer who played a crucial role in the field of extragalactic astronomy and is best known for his discovery of the expanding universe. Born on November 20, 1889, in Marshfield, Missouri, Hubble's early life was marked by an interest in science and sports. His family moved to Chicago, where he excelled in academics and athletics, eventually earning a scholarship to the University of Chicago. There, Hubble initially focused on law and mathematics but was soon drawn to the emerging field of astronomy.

Hubble's academic journey was diverse. After earning a bachelor's degree in mathematics and astronomy, he was awarded a Rhodes Scholarship to attend The Queen's College, Oxford, where he studied jurisprudence and later literature, continuing his broad intellectual pursuits. Despite his initial path in law, Hubble's passion for astronomy persisted. After returning to the United States, he taught high school for a short period before committing himself to astronomical research, obtaining a doctorate from the University of Chicago in 1917. His dissertation, "Photographic Investigations of Faint Nebulae," foreshadowed his future groundbreaking work.

The outbreak of World War I briefly interrupted Hubble's scientific career, as he enlisted in the U.S. Army, serving in France. Following the war, Hubble joined the Mount Wilson Observatory in California, where he would conduct his most influential research. At Mount Wilson, Hubble had access to the 100-inch Hooker Telescope, the world's largest at the time. This powerful instrument allowed him to make observations that would transform our understanding of the universe.

In the 1920s, astronomers were grappling with the nature of so-called "nebulae" and whether they were part of the Milky Way or independent galaxies. Utilizing the Hooker Telescope, Hubble observed Cepheid variable stars within the Andromeda Nebula (now known as the Andromeda Galaxy) and used their periodicity to calculate their distance. His findings, published in 1924, demonstrated that Andromeda lay far outside the Milky Way, establishing it as a separate galaxy. This discovery confirmed the existence of other galaxies, vastly expanding the known scale of the universe and shifting the paradigm from a universe consisting of a single galaxy to one populated with countless galaxies.

Hubble's work did not stop there. He went on to categorize galaxies based on their morphology into a classification system known as the Hubble sequence, or Hubble tuning fork. This system divided galaxies into ellipticals, spirals, and barred spirals, providing a framework for understanding galaxy formation and evolution. Hubble's systematic approach and detailed observations laid the groundwork for modern extragalactic astronomy.

Perhaps Hubble's most profound contribution came in 1929 when he discovered that the universe is expanding. By examining the redshifts of galaxies' spectral lines, he found a relationship between their distances and their velocities, now known as Hubble's Law. This relationship indicated that galaxies are moving away from each other, implying that the universe itself is expanding. Hubble's Law provided the first observational support for the Big Bang theory and revolutionized cosmology, fundamentally altering our understanding of the cosmos.

Hubble's work had far-reaching implications for theoretical physics and cosmology. His discovery of the expanding universe provided crucial evidence for the Big Bang theory, which describes the origin and evolution of the universe. Hubble's observations challenged the then-prevailing static model of the universe and provided a dynamic picture that continues to shape contemporary cosmological theories.

Despite his monumental contributions, Hubble did not receive a Nobel Prize, as the field of astronomy was not included in the prize categories during his lifetime. Nevertheless, his legacy has been immortalized in various ways. In 1990, NASA launched the Hubble Space Telescope, named in his honor. This observatory, orbiting above Earth's atmosphere, has provided unprecedented views of the universe, contributing to countless discoveries and continuing Hubble's legacy of expanding our understanding of the cosmos.

Hubble's personal life was as complex as his scientific endeavors. He married Grace Burke in 1924, and the couple shared a deep interest in the arts and travel. Hubble was known for his charismatic personality and diverse interests, including literature, boxing, and opera. His colleagues and students remembered him as a rigorous scientist with a wide range of intellectual pursuits, reflecting his broad educational background and curiosity.

Throughout his career, Hubble published numerous influential papers and books, including "The Realm of the Nebulae" (1936), which summarized his findings and theories about galaxies and the structure of the universe. His meticulous observational techniques and analytical skills set high standards for astronomical research.

Hubble's impact extended beyond his scientific achievements. He played a significant role in shaping the direction of American astronomy in the mid-20th century. His work inspired a generation of astronomers and astrophysicists, contributing to the establishment of extragalactic astronomy as a major field of study. Hubble's discoveries also had profound philosophical implications, prompting a reevaluation of humanity's place in the universe and the nature of cosmic evolution.

Edwin Hubble continued his research until his death on September 28, 1953. His contributions to astronomy and cosmology have left an indelible mark on science, shaping our understanding of the universe and inspiring future generations to explore the cosmos. His legacy is a testament to the power of observation, curiosity, and the relentless pursuit of knowledge. Through his groundbreaking discoveries, Hubble transformed our view of the universe from a static, limited cosmos to an expansive, dynamic entity, forever altering the course of modern astronomy and cosmology.

49

Chapter 12: Fritz Zwicky

Fritz Zwicky was a Swiss-American astronomer and physicist whose innovative and sometimes controversial ideas significantly shaped modern astrophysics. Born on February 14, 1898, in Varna, Bulgaria, to Swiss parents, Zwicky grew up in Switzerland, where he demonstrated an early aptitude for science and mathematics. He pursued higher education at the Swiss Federal Institute of Technology (ETH Zurich), where he earned a degree in physics in 1920. Zwicky's education in physics provided a solid foundation for his future contributions to astronomy, although his career trajectory would lead him far from his initial field of study.

In 1925, Zwicky moved to the United States to join the California Institute of Technology (Caltech) as a researcher and later as a professor. This move marked the beginning of a long and productive career in astrophysics. At Caltech, Zwicky collaborated with notable scientists, including the renowned astrophysicist Walter Baade. Together, they conducted pioneering research on supernovae, the explosive deaths of stars. In 1934, Zwicky and Baade introduced the concept of neutron stars, dense remnants left after a supernova explosion, significantly advancing our understanding of stellar evolution and the life cycles of stars.

Zwicky is perhaps best known for his contributions to the understanding of dark matter. In 1933, while studying the Coma Cluster of galaxies, he applied the virial theorem to estimate the cluster's mass based on the velocities of its constituent galaxies. To his surprise, he found that the observed mass, inferred from the visible galaxies, was significantly less than the calculated mass required to keep the galaxies bound within the cluster. This discrepancy led Zwicky to propose the existence of what he termed "dunkle Materie" or dark matter, an unseen substance that provides the necessary gravitational pull. This idea was revolutionary and initially met with skepticism, but it has since become a cornerstone of modern cosmology, influencing theories about the structure and evolution of the universe.

Zwicky's contributions to the study of galaxies extended beyond the concept of dark matter. He was a prolific observer and cataloged numerous galaxies, clusters, and supernovae. His comprehensive "Catalogue of Galaxies and Clusters of Galaxies," published in the 1960s, remains a valuable resource for astronomers. Zwicky's observations and classifications laid the groundwork for subsequent large-scale astronomical surveys and provided crucial data for studying the large-scale structure of the universe.

Throughout his career, Zwicky was known for his unconventional thinking and willingness to challenge established ideas. This trait was evident in his development of the "morphological method," an innovative approach to problem-solving that he applied across various fields, including engineering, astronomy, and even social sciences. The morphological method involves systematically exploring all possible solutions to a problem by considering all relevant parameters and their possible combinations. This technique allowed Zwicky to generate novel hypotheses and solutions, contributing to his reputation as a visionary thinker.

Zwicky's contributions to astronomy were not limited to theoretical and observational work; he was also a pioneer in the practical applications of his scientific knowledge. During World War II, he applied his expertise in physics and engineering to develop jet propulsion systems for the U.S. military, significantly advancing the field of aeronautics. His work in this area earned him several patents and recognition for his contributions to the war effort. Zwicky's ability to bridge the gap between theoretical research and practical application was a hallmark of his career.

Despite his many achievements, Zwicky was a polarizing figure within the scientific community. His outspoken and often abrasive personality led to conflicts with colleagues and sometimes hindered the acceptance of his ideas. Nevertheless, his relentless pursuit of truth and his willingness to challenge conventional wisdom left an indelible mark on astrophysics. Zwicky's legacy is reflected in the ongoing research into dark matter, supernovae, and galaxy formation, areas where his pioneering work continues to influence contemporary science.

Zwicky's impact extended beyond his direct contributions to astrophysics. He was a mentor to many students and young researchers, inspiring a new generation of scientists with his passion and innovative thinking. His approach to science, characterized by rigorous observation, bold hypotheses, and a willingness to explore uncharted territories, has become a model for scientific inquiry.

In recognition of his contributions to science, Zwicky received numerous awards and honors throughout his career. These included the Gold Medal of the Royal Astronomical Society and the Presidential Medal of Freedom. Additionally, several astronomical objects and phenomena, including Zwicky clusters and the Zwicky Transient Facility, have been named in his honor, ensuring that his legacy endures in the field of astronomy.

Zwicky's personal life was as dynamic as his professional one. He was married twice and had three daughters. His first marriage to Dorothy Vernon Gates, an artist, ended in divorce. He later married Anna Zurcher, with whom he had a long and happy partnership. Zwicky's personal interests were wide-ranging, encompassing art, philosophy, and mountaineering. He was an avid mountaineer and often combined his love of the outdoors with his scientific pursuits, exploring remote regions and applying his morphological method to various natural phenomena.

Zwicky continued his research and writing until his death on February 8, 1974, in Pasadena, California. His prolific output included numerous papers and several books, including "Morphological Astronomy" and "Discovery, Invention, Research through the

Morphological Approach." These works encapsulate his innovative approach to science and his belief in the power of creative thinking to solve complex problems.

Chapter 13: Subrahmanyan Chandrasekhar

Subrahmanyan Chandrasekhar, one of the most distinguished astrophysicists of the 20th century, was born on October 19, 1910, in Lahore, British India (now Pakistan). His contributions to the field of astrophysics are vast and profound, particularly his work on the theory of stellar structure and evolution. Chandrasekhar's intellectual journey, characterized by rigorous scientific inquiry and a relentless pursuit of excellence, has left an indelible mark on our understanding of the universe.

Chandrasekhar was born into a well-educated family; his father, Chandrasekhara Subrahmanya Ayyar, was a senior officer in the Indian Audits and Accounts Service, and his mother, Sitalakshmi Balakrishnan, was a woman of intellectual acumen, who instilled in him a love for literature and science. He was also the nephew of C.V. Raman, the renowned physicist and Nobel laureate, whose influence played a significant role in shaping Chandrasekhar's scientific career.

From an early age, Chandrasekhar exhibited exceptional mathematical abilities. He completed his early education in Madras (now Chennai), excelling in his studies and demonstrating a particular aptitude for mathematics and physics. At the age of 15, he published his first scientific paper on the cometary theory of stellar evolution, indicating his prodigious talent and early interest in astrophysics.

In 1929, Chandrasekhar was awarded a scholarship to the University of Cambridge, where he pursued his studies under the tutelage of prominent physicists such as Ralph Fowler and Paul Dirac. During his voyage to England, Chandrasekhar worked on a problem that would later become the foundation of his most famous discovery. He considered the fate of white dwarf stars, which are the remnants of stars that have exhausted their nuclear fuel and have shed their outer layers.

Building on the work of Fowler, who had applied the principles of quantum mechanics and the Pauli exclusion principle to white dwarf stars, Chandrasekhar calculated the maximum mass that a white dwarf could have before it would collapse under its own gravity. His calculations revealed that there was a critical mass, now known as the Chandrasekhar limit, approximately 1.4 times the mass of the Sun. Beyond this limit, the electron degeneracy pressure, which supports the white dwarf against gravitational collapse, would be insufficient, leading to the formation of a neutron star or black hole.

Chandrasekhar's groundbreaking work on the Chandrasekhar limit was initially met with skepticism and resistance from some of the leading astronomers of the time, including Sir Arthur Eddington, one of the most influential astrophysicists. Eddington publicly challenged Chandrasekhar's findings, arguing that nature could not allow such a drastic collapse. Despite the controversy and criticism, Chandrasekhar remained steadfast in his convictions, continuing to refine and develop his theories.

In 1933, Chandrasekhar completed his Ph.D. at Cambridge and was elected to a Prize Fellowship at Trinity College. During this period, he further developed his work on stellar dynamics and radiative transfer, laying the groundwork for many of his later contributions to astrophysics. In 1936, he moved to the United States to join the faculty at the University of Chicago, where he would spend the rest of his career and make many of his most significant scientific contributions.

At the University of Chicago, Chandrasekhar joined the Yerkes Observatory, working alongside other eminent scientists such as Otto Struve and William Morgan. His research encompassed a wide range of topics, including the study of stellar atmospheres, the theory of black holes, and hydrodynamic and hydromagnetic stability. One of his notable achievements during this time was his work on the theory of radiative transfer, which describes how radiation is absorbed and scattered by matter. His book "Radiative Transfer," published in 1950, became a classic text in the field and remains influential to this day.

Chandrasekhar's work on the theory of black holes and general relativity also had a profound impact on astrophysics. He developed mathematical solutions to Einstein's equations of general relativity, describing the structure and stability of black holes. His contributions in this area helped to lay the theoretical foundations for understanding these enigmatic objects, which were later confirmed by observational evidence.

In addition to his research, Chandrasekhar was a dedicated and influential teacher. He mentored numerous students who would go on to become leading scientists in their own right. His approach to teaching was characterized by rigorous standards and a deep commitment to nurturing the intellectual growth of his students. Chandrasekhar's ability to convey complex ideas with clarity and precision made him a respected and beloved figure in the academic community.

Chandrasekhar's contributions to astrophysics were recognized with numerous awards and honors. In 1983, he was awarded the Nobel Prize in Physics for his theoretical studies of the physical processes important to the structure and evolution of stars. This recognition was particularly significant as it affirmed the importance and accuracy of his earlier work on the Chandrasekhar limit, which had initially faced considerable opposition.

Throughout his career, Chandrasekhar's work was characterized by a deep sense of curiosity and a relentless pursuit of understanding. He approached scientific problems with a methodical and meticulous mindset, often spending years or even decades on a single topic to achieve a comprehensive and thorough understanding. His contributions extended beyond his own research, as he also made significant efforts to communicate and disseminate scientific knowledge through his writings and lectures.

Chandrasekhar was a prolific author, publishing numerous papers and several influential books. His works include "An Introduction to the Study of Stellar Structure" (1939), "Principles of Stellar Dynamics" (1942), "Hydrodynamic and Hydromagnetic Stability" (1961), and "The Mathematical Theory of Black Holes" (1983). Each of these texts has had a lasting impact on the field of astrophysics, serving as foundational references for generations of researchers.

Despite his many accomplishments, Chandrasekhar remained humble and self-effacing. He often spoke about the importance of perseverance and hard work in scientific research, emphasizing that true understanding requires dedication and patience. His commitment to the pursuit of knowledge and his contributions to science have left a lasting legacy, inspiring countless scientists and researchers around the world.

Chandrasekhar's impact on the field of astrophysics extends beyond his own lifetime. His theories and discoveries continue to shape our understanding of the universe, influencing contemporary research in areas such as stellar evolution, black hole physics, and cosmology. The Chandrasekhar limit, in particular, remains a fundamental concept in astrophysics, providing critical insights into the life cycles of stars and the nature of compact objects.

In recognition of his contributions, numerous honors and memorials have been established in Chandrasekhar's name. These include the Chandra X-ray Observatory, a space telescope launched by NASA in 1999, which has provided invaluable data on black holes, supernova remnants, and other high-energy astrophysical phenomena. Additionally, the Subrahmanyan Chandrasekhar Prize of Plasma Physics, awarded by the Association of Asia Pacific Physical Societies, honors outstanding contributions to plasma physics, reflecting Chandrasekhar's broad impact on various scientific fields.

Chandrasekhar's legacy is also preserved through the many students and colleagues he mentored and inspired. His dedication to fostering intellectual growth and his commitment to scientific excellence have left an enduring mark on the academic community. The principles and values he espoused continue to guide and inspire future generations of scientists.

Chapter 14: Eugene Parker

Eugene Newman Parker was a pioneering American astrophysicist whose groundbreaking work on solar wind and magnetic fields fundamentally changed our understanding of the sun and space weather. Born on June 10, 1927, in Houghton, Michigan, Parker grew up in an intellectually stimulating environment that fostered his interest in science. He completed his undergraduate studies in physics at Michigan State University in 1948 and earned his Ph.D. from the California Institute of Technology (Caltech) in 1951 under the guidance of renowned physicists Carl Anderson and William Fowler. Parker's early education and training at these prestigious institutions equipped him with a strong foundation in theoretical physics and set the stage for his future contributions to astrophysics.

After completing his doctorate, Parker joined the faculty at the University of Utah, where he began his research into plasma physics and astrophysics. In 1955, he moved to the University of Chicago's Enrico Fermi Institute, where he would spend the majority of his career. It was here that Parker conducted his most influential work on the nature of the solar corona and the behavior of charged particles in space.

In the late 1950s, Parker developed the theory of the solar wind, a continuous outflow of charged particles from the sun's corona. This idea was revolutionary at the time, as it challenged the prevailing belief that space between the planets was a vacuum. Parker proposed that the sun's corona, despite being extremely hot, could not be in hydrostatic equilibrium and must be expanding outward, carrying with it a stream of particles. This solar wind, he theorized, would interact with the Earth's magnetosphere and other planetary environments, influencing space weather and geomagnetic storms.

Parker's theory faced significant skepticism from the scientific community. His initial paper on the subject, submitted to the

Astrophysical Journal in 1958, was met with resistance and was nearly rejected. However, it was ultimately published thanks to the support of Subrahmanyan Chandrasekhar, a Nobel Prize-winning astrophysicist and the journal's editor at the time. Chandrasekhar recognized the importance of Parker's work and ensured its publication, despite the controversy it stirred.

The subsequent discovery of the solar wind by spacecraft such as the Soviet Luna 1 and the American Mariner 2 missions in the early 1960s provided empirical validation of Parker's theory. These observations confirmed that the sun continuously emits a stream of charged particles that permeate the solar system, fundamentally altering our understanding of the heliosphere and its interaction with planetary environments. Parker's work laid the foundation for the field of heliophysics, the study of the sun and its effects on the solar system, and opened new avenues for research into space weather and its impacts on Earth.

In addition to his work on the solar wind, Parker made significant contributions to our understanding of magnetic fields in space. He developed the concept of the Parker spiral, which describes the shape of the sun's magnetic field as it extends through the solar system. Due to the rotation of the sun, the solar wind stretches the magnetic field lines into a spiral pattern. This model has been instrumental in explaining the structure of the heliosphere and the behavior of cosmic rays and solar energetic particles.

Parker's research extended beyond solar physics to encompass broader aspects of plasma astrophysics and magnetohydrodynamics (MHD). He investigated the stability of magnetic fields in astrophysical contexts, contributing to the understanding of phenomena such as solar flares, magnetic reconnection, and the dynamics of interstellar and intergalactic media. His theoretical insights into the behavior of plasmas under the influence of magnetic fields have had far-reaching implications for both astrophysics and space exploration.

Throughout his career, Parker published numerous influential papers and books, including "Cosmical Magnetic Fields: Their Origin and Their Activity" (1979), which remains a seminal work in the field. His ability to blend rigorous mathematical analysis with physical intuition allowed him to tackle complex problems and develop theories that have stood the test of time. Parker's work has been recognized with numerous awards and honors, including the National Medal of Science in 1989 and the Kyoto Prize in Basic Sciences in 2003.

Parker's influence extended beyond his scientific contributions. He was a dedicated educator and mentor, inspiring generations of students and researchers at the University of Chicago and beyond. His teaching style emphasized critical thinking, creativity, and the importance of questioning established paradigms. Many of his students went on to become prominent scientists in their own right, carrying forward Parker's legacy of innovation and discovery.

In 2018, Parker's contributions to solar physics were further recognized with the launch of the Parker Solar Probe by NASA. This mission, named in his honor, aims to study the sun's outer corona and the solar wind up close, providing unprecedented data that will deepen our understanding of solar and space physics. The Parker Solar Probe is designed to withstand extreme temperatures and radiation as it approaches the sun's surface, enabling it to collect valuable information about the mechanisms driving the solar wind and the dynamics of the sun's magnetic field. The mission represents a fitting tribute to Parker's pioneering work and his enduring impact on the field of heliophysics.

Parker's personal life was marked by his humility, curiosity, and passion for science. Despite his many accolades, he remained modest about his achievements and focused on the joy of discovery and the pursuit of knowledge. He was known for his warmth and generosity, both as a colleague and a mentor, and his intellectual curiosity extended beyond astrophysics to encompass a wide range of scientific and philosophical interests.

Eugene Parker passed away on March 15, 2022, leaving behind a legacy of groundbreaking discoveries and a profound impact on the field of astrophysics. His work has transformed our understanding of the sun and its interactions with the solar system, shaping the way we study and explore space. Parker's theories continue to guide contemporary research, and his influence is felt in the ongoing quest to unravel the mysteries of the cosmos. His contributions have not only advanced our scientific knowledge but have also inspired a sense of wonder and curiosity about the universe, embodying the spirit of exploration that drives humanity to look beyond our planet and seek a deeper understanding of the natural world.

Chapter 15: Arthur C. Clarke

Arthur C. Clarke was a British science fiction writer, futurist, and inventor whose visionary works and concepts significantly influenced the fields of science fiction and technological innovation. Born on December 16, 1917, in Minehead, Somerset, England, Clarke displayed an early interest in science and astronomy. This fascination persisted throughout his life, driving both his literary and scientific endeavors. He was educated at Huish Grammar School in Taunton before enrolling in King's College London, where he earned a degree in physics and mathematics after World War II.

During the war, Clarke served in the Royal Air Force (RAF) as a radar specialist, working on the development and improvement of radar systems. This experience not only honed his technical skills but also provided him with insights into the potential future applications of technology. After the war, Clarke continued his education and became a prominent member of the British Interplanetary Society, an organization dedicated to promoting space exploration. His involvement with the society further fueled his interest in space and the possibilities it presented.

Clarke's career as a writer began in the late 1930s with the publication of short stories in various science fiction magazines. However, it was his novel "Childhood's End," published in 1953, that established him as a leading figure in science fiction literature. The novel explores themes of human evolution, the potential for extraterrestrial life, and the future of humanity, all of which would become recurring motifs in Clarke's work. "Childhood's End" was notable for its sophisticated treatment of these themes and its ability to blend scientific plausibility with philosophical inquiry, characteristics that would define Clarke's writing style.

Perhaps Clarke's most famous work is "2001: A Space Odyssey," a novel developed in tandem with the 1968 film directed by Stanley

Kubrick. The project began with Clarke and Kubrick collaborating on a screenplay inspired by Clarke's short story "The Sentinel." The result was a groundbreaking film and novel that explored the themes of artificial intelligence, space exploration, and human evolution. "2001: A Space Odyssey" is widely regarded as one of the greatest science fiction films ever made, and Clarke's novelization of the screenplay became a classic in its own right. The story's iconic elements, such as the monolith, the HAL 9000 computer, and the concept of star-child evolution, have left an indelible mark on popular culture and have inspired countless works of science fiction.

Clarke's literary career spanned more than five decades and included numerous novels, short stories, and essays. His "Rama" series, beginning with "Rendezvous with Rama" (1973), is another landmark in science fiction literature. The series explores the discovery of a mysterious alien spacecraft entering the solar system and the subsequent human efforts to explore and understand it. The novels are praised for their detailed scientific realism, imaginative scope, and exploration of humanity's place in the universe. Clarke's ability to combine hard science with imaginative storytelling made his works both intellectually stimulating and entertaining.

In addition to his fiction, Clarke was a prolific writer of non-fiction, where he often explored the future of technology and space exploration. His 1945 paper "Extra-Terrestrial Relays – Can Rocket Stations Give Worldwide Radio Coverage?" proposed the concept of geostationary satellites, a revolutionary idea that laid the groundwork for modern telecommunications. Clarke envisioned a network of satellites orbiting the Earth that could relay signals across the globe, making instantaneous communication possible. This idea was initially met with skepticism, but it eventually became a reality and is now a fundamental component of global communications infrastructure. The geostationary orbit at 35,786 kilometers above the equator is sometimes referred to as the Clarke Orbit in his honor.

Clarke's influence extended beyond his writings to his role as a futurist and public intellectual. He was a frequent commentator on technological and scientific developments, often appearing in interviews, documentaries, and television programs. His predictions about the future, including the advent of personal computers, the internet, and space travel, were remarkably prescient and have earned him a reputation as a visionary thinker. Clarke's three laws of prediction, particularly the third law—"Any sufficiently advanced technology is indistinguishable from magic"—have become widely cited axioms in discussions about technology and its impact on society.

In 1956, Clarke moved to Sri Lanka (then Ceylon), where he spent the remainder of his life. He was drawn to the island by his interest in scuba diving, which he pursued avidly. Clarke's passion for the ocean and its mysteries was evident in his non-fiction work "The Deep Range" and his involvement in marine exploration projects. His move to Sri Lanka also provided him with a unique perspective on global issues, and he became an advocate for environmental conservation and the peaceful use of technology.

Clarke received numerous accolades and honors throughout his career, reflecting his contributions to literature, science, and public understanding of technology. He was knighted in 1998 for his services to literature, and he received several Hugo, Nebula, and Bram Stoker Awards for his science fiction writing. The Arthur C. Clarke Award, established in 1987, is one of the most prestigious awards in science fiction, recognizing outstanding works in the genre.

Despite his achievements, Clarke faced personal challenges, including health issues that affected him in his later years. He was diagnosed with post-polio syndrome in the 1980s, which limited his mobility but did not diminish his intellectual vigor or his passion for writing and exploration. Clarke continued to write and engage with the scientific community until his death on March 19, 2008, at the age of 90. His legacy is preserved not only in his extensive body of work but also in the enduring impact of his ideas on science, technology, and popular culture.

Clarke's vision of the future was one of optimism and possibility, tempered by a recognition of the challenges humanity faces. His works often reflected a belief in the potential for scientific progress to solve problems and improve the human condition. At the same time, he was aware of the ethical and existential questions raised by technological advancement. This dual perspective made his writing particularly resonant, offering both a hopeful vision of the future and a thoughtful consideration of the responsibilities that come with scientific knowledge.

Chapter 16: George Gamow

George Gamow was a towering figure in 20th-century physics and cosmology, known for his pioneering contributions to our understanding of the origins of the universe and the structure of atomic nuclei. Born on March 4, 1904, in Odessa, then part of the Russian Empire, Gamow's intellectual curiosity and innovative thinking propelled him to the forefront of several scientific fields. His work spanned quantum mechanics, nuclear physics, cosmology, and even genetics, making him one of the most versatile and influential scientists of his time.

Gamow's early life was marked by a passion for science and learning. His father was a teacher, and his mother was a musician, providing a nurturing environment that fostered his intellectual growth. Gamow attended the University of Odessa and later the University of Leningrad (now Saint Petersburg State University), where he studied under the guidance of prominent physicists such as Alexander Friedmann and Vladimir Fock. During his time at the University of Leningrad, Gamow became deeply interested in the emerging field of quantum mechanics, which was revolutionizing our understanding of atomic and subatomic processes.

In the late 1920s, Gamow made significant contributions to the field of quantum mechanics. He developed the theory of quantum tunneling, which explained how particles could escape from atomic nuclei, a phenomenon that classical physics could not account for. This work was crucial in explaining alpha decay, a type of radioactive decay in which an atomic nucleus emits an alpha particle (two protons and two neutrons bound together). Gamow's theoretical insights into quantum tunneling laid the groundwork for much of modern nuclear physics and had far-reaching implications for various fields, including chemistry and materials science.

In 1931, Gamow left the Soviet Union and took up positions at several prestigious institutions in Europe, including the University of Copenhagen, where he worked with Niels Bohr, and the University of Cambridge, where he collaborated with Ernest Rutherford. These interactions with leading physicists of the time further honed Gamow's skills and expanded his scientific horizons. In 1934, he moved to the United States, accepting a position at George Washington University in Washington, D.C. It was here that Gamow would make some of his most significant contributions to cosmology and astrophysics.

During the late 1930s and early 1940s, Gamow turned his attention to the origin of the elements and the early universe. He was particularly interested in the process of nucleosynthesis, the formation of chemical elements in the universe. In 1948, Gamow, along with his students Ralph Alpher and Robert Herman, published a groundbreaking paper that laid the foundation for the Big Bang theory. This paper, often referred to as the "$\alpha\beta\gamma$ paper" (pronounced "alpha-beta-gamma" after the authors' initials), proposed that the universe began in a hot, dense state and expanded over time, cooling as it did so. Gamow and his colleagues theorized that the elements observed in the universe today were formed during the first few minutes after the Big Bang, through a process known as primordial nucleosynthesis.

One of the most significant predictions of Gamow's Big Bang model was the existence of a residual heat radiation from the early universe, known today as the cosmic microwave background (CMB). Although Gamow and his collaborators did not detect this radiation themselves, their theoretical work paved the way for its eventual discovery in 1965 by Arno Penzias and Robert Wilson. The discovery of the CMB provided strong evidence for the Big Bang theory and is considered one of the most important observations in cosmology.

Gamow's work extended beyond cosmology to other areas of astrophysics and nuclear physics. He was instrumental in developing the theory of stellar nucleosynthesis, which explains how elements are formed in the cores of stars through nuclear fusion. Gamow's contributions to this field helped to elucidate the processes that power stars and produce the diverse array of elements found in the universe.

In addition to his contributions to physics and cosmology, Gamow had a profound impact on the field of genetics. In the early 1950s, he turned his attention to the problem of how genetic information is encoded in DNA. Inspired by the structure of the DNA molecule, which had been discovered by James Watson and Francis Crick in 1953, Gamow proposed the idea of a genetic code. He suggested that the sequence of nucleotides in DNA could be read in groups of three, known as codons, with each codon corresponding to a specific amino acid. Although Gamow's specific model of the genetic code was not entirely correct, his insights were crucial in guiding subsequent research that ultimately led to the deciphering of the genetic code.

Gamow was also a gifted communicator of science, known for his ability to explain complex concepts in a clear and engaging manner. He wrote numerous popular science books, including the famous "Mr. Tompkins" series, which used a fictional character to explore the principles of modern physics in a whimsical and accessible way. These books introduced generations of readers to the wonders of quantum mechanics, relativity, and cosmology, and remain classics of popular science literature.

Throughout his career, Gamow received numerous honors and awards in recognition of his contributions to science. He was elected to the National Academy of Sciences in 1954 and received the Kalinga Prize for the Popularization of Science from UNESCO in 1956. Despite his many achievements, Gamow remained a humble and approachable figure, known for his playful sense of humor and his enthusiasm for science.

Gamow's legacy is evident in the many areas of physics, cosmology, and biology that he helped to advance. His work on quantum tunneling and nuclear physics laid the foundation for our understanding of radioactive decay and nuclear reactions. His contributions to cosmology, particularly the Big Bang theory and nucleosynthesis, have fundamentally shaped our understanding of the universe's origins and evolution. In genetics, his pioneering ideas about the genetic code provided crucial insights that paved the way for the modern field of molecular biology.

George Gamow passed away on August 19, 1968, in Boulder, Colorado, but his influence on science and his contributions to our understanding of the natural world continue to resonate. His ability to bridge different scientific disciplines and his talent for communicating complex ideas to a broad audience have left an enduring legacy. Gamow's work exemplifies the power of creative thinking and the importance of a broad, interdisciplinary approach to scientific inquiry. His life and achievements serve as an inspiration to scientists and science enthusiasts alike, demonstrating the profound impact that one individual can have on our understanding of the universe.

Chapter 17: Fred Hoyle

Fred Hoyle was a pioneering British astronomer and mathematician whose contributions to astrophysics, particularly in the areas of nucleosynthesis and cosmology, have had a profound impact on our understanding of the universe. Born on June 24, 1915, in Gilstead, a small village near Bingley in West Yorkshire, England, Hoyle grew up in an intellectually stimulating environment. His father was a wool merchant, and his mother was a schoolteacher, both of whom encouraged his early interest in science and mathematics.

Hoyle's formal education began at Bingley Grammar School, where he excelled in mathematics and the sciences. His talents earned him a scholarship to Emmanuel College, Cambridge, in 1933. At Cambridge, he studied under the tutelage of prominent physicists such as Paul Dirac and Arthur Eddington, which profoundly influenced his scientific development. He graduated with a degree in mathematics in 1936 and subsequently pursued postgraduate studies in astrophysics.

During World War II, Hoyle worked for the British Admiralty on radar research, which significantly broadened his technical expertise and problem-solving skills. After the war, he returned to Cambridge, where he began a long and illustrious career in theoretical astrophysics. Hoyle's early work focused on the formation and evolution of stars, leading him to develop groundbreaking theories that would challenge existing paradigms and lay the foundation for future research.

One of Hoyle's most significant contributions to astrophysics was his theory of stellar nucleosynthesis, developed in collaboration with William Fowler and Geoffrey and Margaret Burbidge in the late 1950s. Their seminal paper, known as the B^2FH paper after the initials of the authors, outlined the processes by which elements are formed within stars through nuclear reactions. Hoyle fred his colleagues demonstrated that the elements heavier than hydrogen and helium are synthesized in the interiors of stars through fusion reactions, a process that occurs over the course of stellar lifetimes and during explosive events like supernovae. This work fundamentally changed our understanding of the origin of the elements and the life cycles of stars, earning Fowler the Nobel Prize in Physics in 1983. Although Hoyle did not share the Nobel Prize, his contributions to the field were widely recognized and celebrated.

In addition to his work on nucleosynthesis, Hoyle was a prominent figure in the development of cosmological theories. He is perhaps best known for his advocacy of the steady state theory of the universe, which he developed with Thomas Gold and Hermann Bondi in 1948. The steady state theory posited that the universe is eternal and unchanging on a large scale, with new matter continuously created to maintain a constant density as the universe expands. This theory stood in direct opposition to the Big Bang theory, which proposed that the universe had a definite beginning in a singular explosive event.

Hoyle's support for the steady state theory was partly influenced by his philosophical beliefs about the nature of the universe. He was skeptical of the idea that the universe had a singular beginning, seeing it as reminiscent of religious creation myths. Despite the initial popularity of the steady state theory, accumulating observational evidence, such as the discovery of the cosmic microwave background radiation in 1965, ultimately favored the Big Bang model. Nonetheless, Hoyle's contributions to cosmology were significant, as they spurred further research and debate, enriching our understanding of the universe's origins and evolution.

Hoyle's scientific work was characterized by his boldness and willingness to challenge established ideas. This maverick spirit extended beyond astrophysics to other areas of science. For example, he proposed the controversial panspermia hypothesis, which suggests that life on Earth was seeded by microorganisms from space. Hoyle argued that complex organic molecules, and possibly even life itself, could be distributed throughout the cosmos via comets and meteorites. While this idea remains contentious, it has inspired ongoing research into the origins of life and the potential for extraterrestrial life.

In addition to his scientific research, Hoyle was a prolific writer and communicator of science. He authored numerous books, both technical and popular, aimed at making complex scientific concepts accessible to a broader audience. His popular science books, such as "The Nature of the Universe" (1950) and "Frontiers of Astronomy" (1955), were widely read and appreciated for their clarity and engaging style. Hoyle also wrote several science fiction novels, including "The Black Cloud" (1957) and "A for Andromeda" (1962), which explored scientific themes and speculated about the future of human civilization. His fiction often reflected his deep understanding of science and his imaginative approach to problem-solving.

Hoyle's influence extended beyond his scientific contributions to his role as an educator and mentor. He held various academic positions throughout his career, including Plumian Professor of Astronomy and Experimental Philosophy at the University of Cambridge from 1958 to 1972. He also served as the director of the Institute of Theoretical Astronomy at Cambridge, which he helped establish in 1967. Hoyle's leadership and vision played a crucial role in shaping the institute into a world-renowned center for astrophysical research. His mentorship and collaboration with younger scientists helped foster a new generation of researchers who continued to advance the field.

Despite his many achievements, Hoyle was a controversial figure within the scientific community. His outspoken personality and strong opinions often led to clashes with his peers. He was critical of the scientific establishment and skeptical of the peer review process, which he believed could stifle innovative ideas. Hoyle's willingness to challenge conventional wisdom and pursue unconventional theories sometimes put him at odds with mainstream science, but it also underscored his commitment to intellectual independence and creativity.

Hoyle received numerous honors and awards throughout his career in recognition of his contributions to science. He was elected a Fellow of the Royal Society in 1957 and was knighted in 1972 for his services to astronomy. However, his failure to receive a Nobel Prize, despite his significant contributions to the field, remains a notable omission in the annals of scientific recognition. This oversight is often attributed to his contentious relationships with some members of the scientific community and his unorthodox views.

In his later years, Hoyle continued to write and engage with scientific debates, maintaining his curiosity and enthusiasm for discovery until his death on August 20, 2001. His legacy is preserved not only in his scientific contributions but also in the numerous students and colleagues he inspired throughout his career. Hoyle's work laid the groundwork for much of modern astrophysics and cosmology, and his ideas continue to influence contemporary research.

Hoyle's life and career exemplify the spirit of scientific inquiry and the importance of challenging established paradigms. His bold ideas, whether they were ultimately proven correct or not, pushed the boundaries of our understanding and opened new avenues for exploration. Hoyle's ability to blend rigorous scientific analysis with imaginative speculation made him a unique and influential figure in the history of science.

Chapter 18: Edward Teller

Edward Teller was a Hungarian-American theoretical physicist whose work had a profound impact on nuclear physics and defense policy during the 20th century. Born on January 15, 1908, in Budapest, Austria-Hungary, Teller grew up in a well-educated and affluent Jewish family. His father, Max Teller, was a lawyer, and his mother, Ilona Deutsch, was a pianist. Teller's early education emphasized science and mathematics, subjects in which he excelled. He showed an early aptitude for physics and was influenced by the political and scientific environment of post-World War I Europe.

Teller's academic journey began at the University of Karlsruhe in Germany, where he studied chemical engineering. However, his true passion lay in physics, and he transferred to the University of Leipzig to study under Werner Heisenberg, one of the pioneers of quantum mechanics. Teller earned his Ph.D. in physics in 1930 with a dissertation on molecular spectra. His early research focused on quantum mechanics, particularly on the hydrogen molecular ion, which laid the groundwork for his later contributions to nuclear physics.

The rise of the Nazi regime in Germany forced Teller, who was Jewish, to flee Europe in the 1930s. He moved to the United States in 1935, where he joined the faculty at George Washington University. There, he worked with George Gamow, a prominent physicist known for his work on stellar nucleosynthesis and cosmology. During this period, Teller's research interests shifted towards nuclear physics, particularly the potential of nuclear fission as a source of energy.

With the outbreak of World War II and the subsequent discovery of nuclear fission, Teller became involved in the U.S. atomic bomb project. He was one of the key scientists recruited for the Manhattan Project, the secret U.S. government effort to develop atomic weapons. Teller worked at Los Alamos National Laboratory in New Mexico, where he played a crucial role in the theoretical development of the bomb. Although the primary focus was on the creation of fission bombs, Teller was already envisioning a more powerful weapon: the hydrogen bomb, or thermonuclear bomb.

Teller's advocacy for the hydrogen bomb set him apart from many of his colleagues. He argued that the potential for a much more powerful weapon justified its development, despite the significant technical challenges involved. After the successful detonation of the atomic bombs over Hiroshima and Nagasaki in 1945, Teller intensified his efforts to develop a thermonuclear weapon. His persistence paid off when the first successful test of a hydrogen bomb, known as "Ivy Mike," was conducted on November 1, 1952, at Enewetak Atoll in the Pacific Ocean. This test confirmed that thermonuclear weapons were not only feasible but vastly more powerful than their fission-based predecessors.

Teller's role in the development of the hydrogen bomb earned him the moniker "the father of the hydrogen bomb," a title he shared with other key figures such as Stanislaw Ulam and Enrico Fermi. The success of the hydrogen bomb had significant implications for global geopolitics, ushering in an era of nuclear arms races and Cold War tensions. Teller's advocacy for nuclear weapons and his involvement in defense policy extended beyond their development. He was a staunch proponent of maintaining a robust nuclear arsenal as a deterrent against potential adversaries, particularly the Soviet Union.

Teller's views on nuclear weapons and defense policy often brought him into conflict with other scientists and policymakers. One of the most notable controversies of his career was his role in the hearings that led to the revocation of J. Robert Oppenheimer's security clearance in 1954. Oppenheimer, who had been the scientific director of the Manhattan Project, faced accusations of disloyalty and Communist sympathies. Teller testified against Oppenheimer, expressing doubts about his loyalty and judgment. This testimony was pivotal in the decision to revoke Oppenheimer's clearance and effectively ended

Oppenheimer's career in government science. Teller's actions during the hearings deeply divided the scientific community, and many of his colleagues viewed him as a traitor to the cause of scientific integrity and cooperation.

Despite the controversies, Teller's contributions to science and technology were substantial. In addition to his work on nuclear weapons, he made significant contributions to theoretical physics, particularly in the areas of quantum mechanics, statistical mechanics, and plasma physics. Teller's research on the hydrogen molecular ion and his development of the Jahn-Teller effect, which describes the geometric distortion of molecules in certain situations, are particularly notable. His work on the Monte Carlo method, a statistical technique for solving physical and mathematical problems, has had wide-ranging applications in various fields of science and engineering.

Teller was also a passionate advocate for the peaceful uses of nuclear energy. He believed that nuclear power could provide a virtually limitless source of energy and argued for its development as an alternative to fossil fuels. Teller's vision extended to ambitious projects such as the use of nuclear explosions for large-scale engineering projects, a concept known as "Project Plowshare." Although many of these ideas were not realized, they reflected Teller's belief in the transformative potential of nuclear technology.

In the latter part of his career, Teller became increasingly involved in national security and defense policy. He was a key proponent of the Strategic Defense Initiative (SDI), a proposed missile defense system championed by President Ronald Reagan in the 1980s. The SDI, often referred to as "Star Wars," aimed to develop a space-based missile defense system to protect the United States from nuclear attacks. Teller's advocacy for the SDI was consistent with his belief in the importance of technological superiority as a means of ensuring national security.

Teller's legacy is complex and multifaceted. On one hand, he was a brilliant scientist whose contributions to nuclear physics and defense policy have had a lasting impact. His work on the hydrogen bomb and his advocacy for a strong nuclear deterrent played a crucial role in shaping the strategic landscape of the Cold War. On the other hand, his actions during the Oppenheimer hearings and his unwavering support for nuclear weapons made him a controversial figure. Teller's willingness to challenge prevailing norms and his uncompromising stance on national security issues earned him both admiration and criticism.

Throughout his life, Teller received numerous honors and awards in recognition of his scientific achievements. He was elected to the National Academy of Sciences in 1948 and received the Enrico Fermi Award in 1962 for his contributions to nuclear science. In 2003, he was awarded the Presidential Medal of Freedom, the highest civilian honor in the United States, in recognition of his contributions to science and national security.

Teller's personal life was marked by his marriage to Augusta Maria Harkanyi, a fellow Hungarian émigré, in 1934. The couple had two children, a son, Paul, and a daughter, Wendy. Augusta, who was also a scientist, provided a supportive partnership throughout Teller's career. Teller was known for his strong personality, his intellectual rigor, and his dedication to his work. He was deeply committed to the idea that science and technology could be harnessed for the betterment of humanity, even as he recognized the potential dangers they posed.

Edward Teller passed away on September 9, 2003, in Stanford, California, at the age of 95. His death marked the end of a remarkable career that spanned some of the most momentous events in modern history. Teller's contributions to nuclear physics, his role in the development of nuclear weapons, and his advocacy for national security have left an indelible mark on the 20th century. His legacy continues to be the subject of debate and reflection, as scientists, policymakers, and historians grapple with the implications of his work and the ethical questions it raises.

Chapter 19: Allan Sandage

Allan Rex Sandage was a pioneering American astronomer whose work fundamentally shaped our understanding of the universe's scale, structure, and age. Born on June 18, 1926, in Iowa City, Iowa, Sandage's contributions to observational cosmology, particularly his measurements of the Hubble constant and the expansion rate of the universe, have left a lasting legacy in the field of astrophysics. Over a career spanning more than half a century, Sandage's meticulous observations and analyses provided critical insights into the life cycles of stars, the distances to galaxies, and the age and size of the universe.

Sandage's early life was marked by an interest in science and engineering. He grew up in a supportive family that encouraged his intellectual pursuits. After completing high school, Sandage attended the University of Illinois, where he initially studied engineering before switching to physics. His education was interrupted by World War II, during which he served in the U.S. Navy. After the war, he resumed his studies and graduated with a bachelor's degree in physics in 1948. Sandage then went on to pursue graduate studies in astronomy at the California Institute of Technology (Caltech), where he studied under the renowned astronomer Walter Baade. He completed his Ph.D. in 1953 with a dissertation on the light curves of variable stars, which set the stage for his future research in observational astronomy.

After earning his doctorate, Sandage joined the staff of the Mount Wilson and Palomar Observatories, where he began a fruitful collaboration with Edwin Hubble, the astronomer famous for discovering the expanding universe. Sandage's work with Hubble had a profound impact on his career and set the trajectory for his research in cosmology. Following Hubble's death in 1953, Sandage took on the monumental task of continuing and expanding Hubble's observational program. He focused on refining the measurements of the Hubble constant, the rate at which the universe is expanding, which is crucial for determining the size and age of the universe.

One of Sandage's most significant contributions to astronomy was his work on the distance scale of the universe. He used the 200-inch Hale Telescope at Palomar Observatory to observe and measure the distances to galaxies. Sandage employed several methods to determine these distances, including the use of Cepheid variable stars, which are stars that pulsate in a predictable manner and can be used as "standard candles" to measure astronomical distances. By identifying Cepheid variables in nearby galaxies, Sandage was able to calibrate the distance scale and refine the value of the Hubble constant. His meticulous observations and analyses led to more accurate measurements of the distances to galaxies and a better understanding of the expansion rate of the universe.

Sandage's work on the Hubble constant was not without controversy. His early measurements suggested a value for the Hubble constant that implied a much older universe than previous estimates. This discrepancy sparked a lively debate within the astronomical community and prompted further observations and theoretical developments. Over the years, Sandage continued to refine his measurements, incorporating new data and methods as they became available. His efforts contributed to a more precise determination of the Hubble constant, which is still a topic of active research and debate today.

In addition to his work on the Hubble constant, Sandage made significant contributions to the study of stellar populations and the life cycles of stars. He was particularly interested in understanding the ages of the oldest stars in the universe, which provide important clues about the age of the universe itself. Sandage's observations of globular clusters, which are dense groups of old stars, helped to establish the ages of these stellar systems and provided evidence for the formation and evolution of the Milky Way galaxy.

Sandage also played a key role in the study of quasars, which are extremely luminous and distant objects powered by supermassive black holes at the centers of galaxies. In the 1960s, Sandage and his colleagues discovered several quasars and conducted detailed studies of their properties. These observations provided critical insights into the nature of quasars and their role in the evolution of galaxies.

Throughout his career, Sandage was known for his meticulous observational techniques and his dedication to precision in astronomical measurements. He was a master of using large telescopes and advanced instrumentation to gather high-quality data. His careful and methodical approach to data analysis set a high standard for observational astronomy and earned him a reputation as one of the foremost astronomers of his time.

Sandage's contributions to astronomy were widely recognized, and he received numerous honors and awards for his work. He was elected to the National Academy of Sciences in 1967 and received the Gold Medal of the Royal Astronomical Society in 1967, one of the highest honors in the field of astronomy. In 1970, he was awarded the prestigious Crafoord Prize by the Royal Swedish Academy of Sciences, often considered the equivalent of the Nobel Prize for fields not covered by the Nobel awards. Sandage also received the Bruce Medal from the Astronomical Society of the Pacific in 1975 and the National Medal of Science in 1973.

In addition to his scientific achievements, Sandage was a prolific writer and communicator of science. He authored or co-authored over 500 scientific papers and several influential books, including "The Hubble Atlas of Galaxies," which remains a key reference work in the field of galaxy morphology. Sandage's ability to explain complex astronomical concepts in a clear and accessible manner made his writings valuable resources for both professional astronomers and the general public.

Sandage's personal life was marked by a deep commitment to his work and a passion for discovery. He was known for his intense focus and dedication, often spending long hours at the observatory or in front of his computer analyzing data. Despite his demanding schedule, Sandage was a supportive mentor to many young astronomers, providing guidance and encouragement to the next generation of researchers.

Sandage's influence extended beyond his scientific contributions. He was a vocal advocate for the importance of observational astronomy and the need for continued investment in large telescopes and advanced instrumentation. His efforts helped to secure funding for major observatories and research programs, ensuring that future generations of astronomers would have the tools they needed to continue exploring the universe.

In his later years, Sandage continued to be active in research and maintained a keen interest in the latest developments in astronomy. He remained a prominent figure in the astronomical community until his death on November 13, 2010, at the age of 84. Sandage's legacy lives on through the many discoveries and advancements he made during his remarkable career, and his work continues to inspire astronomers and scientists around the world.

Chapter 20: William Pickering

William Hayward Pickering, a pivotal figure in space exploration, was born on December 24, 1910, in Wellington, New Zealand. He moved to the United States in 1929 to study electrical engineering at the California Institute of Technology (Caltech), where he completed his bachelor's degree in 1932 and his master's degree in 1933. Pickering's journey into space science began during his tenure as a student and later as a professor at Caltech. His early work focused on cosmic rays, using balloon-borne instruments to measure their intensity and distribution in the upper atmosphere.

Pickering's career took a significant turn when he joined the Jet Propulsion Laboratory (JPL) in Pasadena, California, during World War II. Initially, JPL was heavily involved in developing rocket technology for military purposes, but Pickering's interest was always in using these rockets for scientific exploration. After the war, he played a crucial role in transitioning JPL from a military research facility to a center for space exploration. His leadership and vision were instrumental in shaping the lab's future direction.

In 1954, Pickering was appointed Director of JPL. His tenure marked a period of rapid advancement in the United States' space capabilities. One of his first major projects was the Explorer program, which aimed to launch the first American satellite into space. Under Pickering's guidance, JPL successfully launched Explorer 1 on January 31, 1958. This satellite made a groundbreaking discovery by detecting the Van Allen radiation belts, regions of charged particles trapped by Earth's magnetic field. This achievement was a significant milestone in the Space Race, marking the United States' entry into space exploration.

Pickering's leadership extended beyond satellite launches. He was instrumental in the development of the Pioneer and Ranger programs, which sent probes to study the Moon and other planets. The Ranger missions, despite initial failures, ultimately succeeded in providing the first close-up images of the lunar surface, paving the way for the Apollo missions. Pickering's commitment to overcoming setbacks and his emphasis on learning from failures were key factors in these successes.

The Mariner program, another of Pickering's significant contributions, further demonstrated his vision and expertise. The Mariner missions were a series of interplanetary probes that provided detailed information about Venus, Mars, and Mercury. Mariner 2, launched in 1962, was the first successful mission to another planet, conducting a flyby of Venus and sending back valuable data about its atmosphere and surface conditions. Mariner 4, which flew by Mars in 1965, provided the first close-up images of the Martian surface, revealing a landscape with craters and a thin atmosphere, vastly different from Earth.

Pickering's influence extended to the broader field of planetary science. He advocated for the importance of robotic exploration as a precursor to human missions, a philosophy that continues to guide space exploration strategies today. His work laid the foundation for subsequent missions, including the Voyager program, which sent probes to the outer planets, and the Mars rovers, which have revolutionized our understanding of the Red Planet.

In addition to his technical contributions, Pickering was a passionate advocate for space science education and public outreach. He understood the importance of inspiring the next generation of scientists and engineers. Throughout his career, he gave numerous lectures and public talks, explaining the significance of space exploration and its benefits to society. He also emphasized international cooperation in space, believing that the exploration of space should be a global endeavor.

Pickering's achievements earned him numerous accolades and honors. He was awarded the National Medal of Science in 1975 by President Gerald Ford, recognizing his contributions to the development of space technology and his leadership in space exploration. He was also inducted into the International Space Hall of Fame and received honorary degrees from several universities.

Despite his many accomplishments, Pickering remained a humble and dedicated scientist. His colleagues admired his ability to remain calm under pressure and his unwavering commitment to excellence. He was known for his hands-on approach, often spending long hours in the laboratory and personally overseeing the progress of various projects. His leadership style was characterized by a collaborative approach, encouraging teamwork and innovation among his staff.

Pickering retired from JPL in 1976 but continued to be involved in space science and education. He served on advisory panels and committees, offering his expertise to guide the future direction of space exploration. He also worked with international organizations to promote space research and development in other countries.

William Pickering passed away on March 15, 2004, leaving behind a legacy of pioneering achievements in space exploration. His contributions laid the groundwork for many of the advancements in space science that followed. His vision of using robotic missions to explore the solar system has been realized through numerous successful missions, and his emphasis on international collaboration continues to shape global space policies. Pickering's life and work remain an inspiration to scientists and engineers around the world, demonstrating the power of perseverance, innovation, and dedication in the pursuit of knowledge.

Chapter 21: Fred Lawrence Whipple

Fred Lawrence Whipple, born on November 5, 1906, in Red Oak, Iowa, was a groundbreaking American astronomer whose innovative ideas significantly advanced our understanding of comets and small bodies in the solar system. Whipple's journey into the field of astronomy began at an early age, fueled by his fascination with the night sky and the celestial bodies that populate it. He pursued his undergraduate studies at Occidental College in Los Angeles, where he initially focused on mathematics, but his passion for astronomy soon took precedence.

In 1927, Whipple enrolled in the University of California, Berkeley, to pursue graduate studies in astronomy. He earned his Ph.D. in 1931 under the supervision of Harold F. Weaver, with a dissertation that focused on the study of meteors. During his time at Berkeley, Whipple developed a keen interest in the behavior and composition of meteors, which would later influence his groundbreaking theories about comets.

After completing his doctorate, Whipple joined the Harvard College Observatory in 1931, where he would spend much of his illustrious career. His early work at Harvard involved studying meteors and their interactions with Earth's atmosphere. He developed methods to photograph meteors and calculate their orbits, contributing to a deeper understanding of these transient celestial phenomena. Whipple's meticulous observations and calculations laid the groundwork for his later theories on the nature of comets.

One of Whipple's most significant contributions to astronomy came in the early 1950s with his proposal of the "dirty snowball" model of comets. Prior to Whipple's theory, the nature of comets was poorly understood, with prevailing models unable to adequately explain their behavior. Whipple proposed that comets were composed of a mixture of volatile ices—such as water, carbon dioxide, and methane—mixed with dust and rocky material. This "dirty snowball" model explained how comets could produce glowing comas and tails when they approached the Sun. The heat from the Sun would cause the ices to sublimate, releasing gas and dust into space, which would then be illuminated by sunlight and interact with the solar wind to form the characteristic tail.

Whipple's model was revolutionary because it provided a cohesive explanation for the observed properties of comets, including their orbits, physical structure, and behavior near the Sun. His theory was initially met with skepticism, but it was eventually confirmed through observations and space missions. The success of the model transformed the understanding of comets and established Whipple as a leading figure in planetary science.

In addition to his theoretical work, Whipple made significant contributions to practical astronomy and space science. During World War II, he worked on the development of radar countermeasures and contributed to the creation of chaff, a radar jamming device used to protect aircraft from detection. This work earned him the Presidential Certificate of Merit in 1948. His experience with radar technology also influenced his later work in tracking artificial satellites.

In the late 1950s, Whipple played a crucial role in the development of the United States' satellite tracking network. As the director of the Smithsonian Astrophysical Observatory (SAO), he oversaw the creation of the Moonwatch program, a network of amateur astronomers who tracked artificial satellites. This program was instrumental in tracking the Soviet Union's Sputnik satellites, providing valuable data on their orbits and contributing to the early days of the space race.

Under Whipple's leadership, the SAO expanded its research efforts and became a leading institution in space science. He spearheaded the development of the Baker-Nunn camera, a highly precise wide-angle camera designed to track satellites and other celestial objects. The data collected by these cameras were crucial for understanding satellite orbits and improving the accuracy of astronomical observations.

Whipple's work extended beyond comets and satellite tracking. He also made significant contributions to the study of meteoroids and the interplanetary medium. He developed the concept of the "Whipple shield," a protective layer designed to shield spacecraft from micrometeoroid impacts. This concept has been widely adopted in spacecraft design and is a testament to Whipple's innovative thinking and his ability to apply his knowledge of celestial phenomena to practical engineering challenges.

Throughout his career, Whipple received numerous accolades and honors for his contributions to astronomy and space science. He was elected to the National Academy of Sciences in 1959 and received the Bruce Medal of the Astronomical Society of the Pacific in 1983 for his lifetime contributions to astronomy. He was also awarded the Gold Medal of the Royal Astronomical Society in 1986, one of the highest honors in the field.

In addition to his scientific achievements, Whipple was a dedicated educator and mentor. He taught at Harvard University for many years, inspiring generations of students with his enthusiasm for astronomy and his deep knowledge of the subject. His ability to communicate complex scientific ideas in an accessible manner made him a beloved figure among his students and colleagues.

Whipple's legacy extends beyond his scientific contributions. His pioneering work laid the foundation for future space missions, including those that have visited and studied comets up close. The European Space Agency's Rosetta mission, which successfully landed a probe on Comet 67P/Churyumov-Gerasimenko in 2014, was built on the principles established by Whipple's "dirty snowball" model. This mission provided detailed observations that confirmed many aspects of Whipple's theory and advanced our understanding of cometary composition and behavior.

Whipple's influence is also evident in the ongoing efforts to track and study near-Earth objects (NEOs). His early work on meteor and satellite tracking paved the way for modern NEO tracking programs, which are essential for assessing the potential threat of asteroid impacts and protecting our planet from these hazards.

Fred Lawrence Whipple passed away on August 30, 2004, leaving behind a rich legacy of scientific discovery and innovation. His contributions to astronomy and space science have had a lasting impact, shaping our understanding of comets, meteors, and the interplanetary medium. Whipple's work exemplifies the power of curiosity, creativity, and perseverance in the pursuit of knowledge. His legacy continues to inspire astronomers, scientists, and engineers to explore the mysteries of the universe and push the boundaries of what we know about the cosmos.

Chapter 22: John Glenn

John Herschel Glenn Jr., born on July 18, 1921, in Cambridge, Ohio, was an iconic American aviator, astronaut, and politician whose life and career were marked by groundbreaking achievements and a profound dedication to public service. Glenn's journey into the annals of history began with his early fascination with flight and exploration. Raised in New Concord, Ohio, Glenn's interest in aviation was sparked at an early age by barnstorming pilots and his father's tales of World War I.

Glenn attended Muskingum College, where he studied engineering, but his education was interrupted by the outbreak of World War II. Driven by a sense of duty and adventure, he enlisted in the Naval Aviation Cadet Program in 1942, eventually earning his wings as a Marine Corps aviator. During the war, Glenn flew 59 combat missions in the South Pacific, demonstrating exceptional skill and bravery. His wartime service earned him two Distinguished Flying Crosses and ten Air Medals, laying the foundation for his reputation as a fearless and capable pilot.

After World War II, Glenn remained in the military and continued to distinguish himself as a pilot. He served in the Korean War, flying 90 combat missions and earning additional accolades, including eight more Air Medals and two more Distinguished Flying Crosses. His combat experience honed his skills and solidified his status as one of the Marine Corps' top aviators. It was during this time that Glenn began to garner attention for his exceptional flying abilities and his cool demeanor under pressure.

In 1954, Glenn attended the U.S. Naval Test Pilot School at Patuxent River, Maryland, where he further developed his expertise in high-speed flight. As a test pilot, he set a transcontinental speed record in 1957, flying an F8U Crusader from Los Angeles to New York in just 3 hours and 23 minutes. This achievement, known as "Project Bullet," captured the nation's imagination and underscored Glenn's status as a pioneering aviator.

Glenn's prowess as a pilot and his exemplary service record made him a prime candidate for the burgeoning space program. In 1959, he was selected as one of the original seven astronauts for NASA's Project Mercury, the United States' first human spaceflight program. The selection process was rigorous, and Glenn's military background, engineering education, and test pilot experience made him an ideal choice. Alongside his fellow "Mercury Seven" astronauts, Glenn became a national hero, symbolizing the United States' aspirations in the space race against the Soviet Union.

On February 20, 1962, John Glenn made history as the first American to orbit the Earth aboard the spacecraft Friendship 7. The mission, designated MA-6, was a pivotal moment in the space race and a major milestone for NASA. During the nearly five-hour flight, Glenn orbited the Earth three times, demonstrating the feasibility of human spaceflight and the effectiveness of NASA's Mercury spacecraft. The mission was not without challenges, including a potentially catastrophic heat shield problem, but Glenn's calm and methodical approach ensured its success. His safe return to Earth made him an instant national hero and a symbol of American ingenuity and determination.

Glenn's achievement was celebrated around the world, and he received numerous honors and accolades. He was awarded the NASA Distinguished Service Medal by President John F. Kennedy and embarked on a goodwill tour that took him to 24 countries. Glenn's heroism and charisma made him a beloved figure, and he became a powerful advocate for the space program and the broader scientific and technological advancements it represented.

Following his historic flight, Glenn continued to serve NASA and the United States, but his desire to contribute in new ways led him to pursue a career in politics. He retired from the Marine Corps as a colonel in 1965 and ran for the U.S. Senate from Ohio. Although his initial bid in 1964 was thwarted by a serious injury, Glenn remained undeterred and successfully won a Senate seat in 1974. As a senator, he served for 24 years, from 1974 to 1999, representing Ohio as a Democrat.

During his tenure in the Senate, Glenn focused on issues related to science, technology, and national security. He was a staunch advocate for the space program and played a key role in shaping space policy. He also worked on environmental protection, nuclear non-proliferation, and government ethics reform. Glenn's legislative efforts reflected his deep commitment to public service and his belief in the transformative power of science and technology.

In 1998, at the age of 77, Glenn made a historic return to space as a crew member on the Space Shuttle Discovery's STS-95 mission. This mission made him the oldest person to fly in space and provided valuable data on the effects of spaceflight on the aging body. Glenn's participation in the mission was a testament to his enduring spirit of exploration and his dedication to advancing human knowledge. His return to space captivated the public and reaffirmed his status as an American icon.

Throughout his life, Glenn was known for his humility, integrity, and unwavering commitment to his country. He was awarded the Presidential Medal of Freedom in 2012, the highest civilian honor in the United States, in recognition of his contributions to aviation, space exploration, and public service. Glenn's legacy is also enshrined in numerous institutions and honors, including the John Glenn College of Public Affairs at Ohio State University and the John Glenn Research Center at NASA.

John Glenn passed away on December 8, 2016, at the age of 95. His death marked the end of an era, but his legacy lives on through the countless lives he touched and the enduring impact of his achievements. Glenn's life story is a testament to the power of courage, perseverance, and a relentless pursuit of excellence. From his early days as a combat pilot to his groundbreaking spaceflights and distinguished political career, Glenn exemplified the best of American values and inspired generations to reach for the stars.

His contributions to space exploration, in particular, have had a lasting impact on humanity's quest to explore the cosmos. Glenn's historic orbital flight helped pave the way for subsequent manned missions, including the Apollo moon landings, and his advocacy for the space program ensured continued support for NASA's missions. His return to space on the Space Shuttle Discovery underscored the importance of lifelong learning and the potential for human achievement at any age.

In addition to his professional accomplishments, Glenn's personal qualities endeared him to many. He was known for his modesty, often deflecting praise to his colleagues and the teams that supported him. His dedication to his family, particularly his wife Annie, whom he married in 1943, was a cornerstone of his life. The couple's enduring partnership, marked by mutual support and resilience, was a source of inspiration to many.

John Glenn's legacy extends beyond his individual achievements. He represents the spirit of exploration and the drive to push beyond known boundaries in the pursuit of knowledge and progress. His life story continues to inspire new generations of scientists, engineers, and explorers to dream big and pursue their goals with determination and integrity. As we continue to explore the frontiers of space, John Glenn's contributions and his enduring spirit of adventure will always be remembered as foundational to the journey.

Chapter 23: Richard Feynman

Richard Phillips Feynman, born on May 11, 1918, in Queens, New York, was a renowned American theoretical physicist whose contributions to quantum mechanics, quantum electrodynamics, and particle physics earned him a Nobel Prize and a lasting legacy in the world of science. Feynman's journey into the realm of physics began early, influenced by his father's encouragement to question and understand the world around him. His mother, meanwhile, instilled in him a sense of humor and a lively spirit, traits that would become as characteristic of Feynman as his intellectual brilliance.

Feynman's prodigious talent for mathematics was evident from a young age. He taught himself advanced mathematics and created his own symbols to represent different operations. After excelling in high school, he attended the Massachusetts Institute of Technology (MIT), where he earned a bachelor's degree in physics in 1939. His time at MIT was marked by his intense curiosity and unconventional problem-solving approaches, which often involved deep, intuitive understanding rather than rote memorization.

Following MIT, Feynman pursued his Ph.D. at Princeton University, where he studied under John Archibald Wheeler. His doctoral thesis on the "Principle of Least Action in Quantum Mechanics" laid the groundwork for his future contributions to quantum electrodynamics (QED). During this period, Feynman developed a reputation for his unconventional thinking and his ability to tackle complex problems with innovative solutions. He also began to develop the path integral formulation of quantum mechanics, a novel approach that would later become a cornerstone of modern theoretical physics.

World War II interrupted Feynman's academic pursuits, and he joined the Manhattan Project, the secret U.S. effort to develop the atomic bomb. At Los Alamos National Laboratory, Feynman worked on critical issues related to the bomb's design, particularly in the area of neutron diffusion. His work was crucial to the project's success, and he was known for his unorthodox methods, such as his habit of cracking safes containing classified information as a way to highlight security flaws. Despite the gravity of the work, Feynman's playful spirit and knack for problem-solving endeared him to his colleagues.

After the war, Feynman returned to academia, taking a position at Cornell University. It was during his time at Cornell that he made some of his most significant contributions to physics. He developed the Feynman diagrams, a revolutionary pictorial representation of the behavior of subatomic particles that simplified complex calculations in quantum electrodynamics. These diagrams provided a visual tool to understand and compute interactions between particles, making it easier for physicists to predict the outcomes of particle collisions and other quantum phenomena. Feynman diagrams became an essential part of the toolkit for physicists and remain widely used today.

In 1950, Feynman moved to the California Institute of Technology (Caltech), where he spent the rest of his career. At Caltech, he continued to make groundbreaking contributions to theoretical physics, including his work on the theory of superfluidity in liquid helium and his contributions to the development of quantum computing and nanotechnology. His research was characterized by deep originality and a willingness to challenge established ideas.

One of Feynman's most notable achievements was his contribution to quantum electrodynamics (QED), the theory that describes how light and matter interact. Along with Julian Schwinger and Sin-Itiro Tomonaga, Feynman developed a comprehensive formulation of QED that resolved inconsistencies in earlier theories. For this work, the trio was awarded the Nobel Prize in Physics in 1965. Feynman's approach, which included his innovative use of Feynman diagrams, provided a powerful framework for understanding the interactions of photons and electrons, fundamentally shaping the field of particle physics.

Feynman's impact on science extended beyond his research. He was a passionate and effective educator, known for his ability to communicate complex ideas in a clear and engaging manner. His lectures at Caltech were legendary, and his ability to simplify and explain intricate concepts inspired generations of students. His series of lectures, later compiled into the "Feynman Lectures on Physics," remains a seminal resource in physics education, blending rigorous scientific analysis with Feynman's unique pedagogical style.

Feynman's curiosity and love of learning extended beyond physics. He was an avid bongo drummer, a talented artist, and an enthusiastic safe-cracker. He explored diverse interests ranging from biology to art, often bringing insights from one field into another. His book "Surely You're Joking, Mr. Feynman!"—a collection of autobiographical anecdotes—reveals his multifaceted personality and his insatiable curiosity about the world. The book became a bestseller and introduced Feynman to a broader audience, showcasing his wit, irreverence, and relentless pursuit of knowledge.

In 1986, Feynman played a crucial role in the investigation of the Space Shuttle Challenger disaster. As a member of the Rogers Commission, he brought his trademark rigor and skepticism to the inquiry. Feynman's famous demonstration—dropping a rubber O-ring into a glass of ice water to show how it lost its elasticity in cold temperatures—highlighted the flaw that led to the disaster. His insistence on transparency and accountability underscored his commitment to scientific integrity and public safety.

Feynman's later years were marked by his continued contributions to science and education, despite battling cancer. He remained active in research, exploring new areas such as quantum computing and contributing to the development of the field of nanotechnology. His final book, "What Do You Care What Other People Think?"—another collection of personal anecdotes and reflections—was published in

1988 and provided further insight into his philosophical outlook and his approach to life and science.

Richard Feynman passed away on February 15, 1988, leaving behind a legacy of scientific innovation, pedagogical excellence, and an enduring influence on the culture of physics. His approach to science—characterized by curiosity, creativity, and a willingness to challenge conventional wisdom—continues to inspire scientists and educators. Feynman's contributions to quantum mechanics and electrodynamics have had a lasting impact on theoretical physics, shaping our understanding of the fundamental forces of nature.

Beyond his scientific achievements, Feynman's life serves as a testament to the joy of discovery and the importance of questioning and exploring the world. His ability to blend deep intellectual rigor with a playful and irreverent spirit made him a beloved figure both within and outside the scientific community. Feynman's legacy endures in the many students he inspired, the colleagues he influenced, and the countless individuals who continue to learn from his writings and teachings.

Chapter 24: Clyde Tombaugh

Clyde William Tombaugh, born on February 4, 1906, in Streator, Illinois, was an American astronomer best known for his discovery of Pluto in 1930, which marked a significant milestone in the field of planetary science. Tombaugh's life and career were characterized by his passion for astronomy, meticulous observational skills, and a relentless pursuit of knowledge that led to one of the most celebrated discoveries in the history of astronomy.

Tombaugh's early years were spent on a farm in Kansas, where his family moved when he was a child. Growing up in a rural environment, Tombaugh developed a keen interest in the night sky. His fascination with astronomy was sparked by the clear, star-filled skies of the Kansas plains and further fueled by his reading of popular science books and magazines. Despite the financial constraints of his family, Tombaugh's parents supported his interest in science, recognizing his talent and enthusiasm.

After graduating from high school, Tombaugh's plans to attend college were derailed by a devastating hailstorm that destroyed his family's crops, leaving them unable to afford tuition. Undeterred, Tombaugh pursued his passion for astronomy on his own. Using spare parts from farm equipment and simple materials, he built his own telescopes, demonstrating remarkable ingenuity and resourcefulness. His homemade telescopes allowed him to make detailed observations of the planets, which he meticulously recorded and sketched.

Tombaugh's precise and detailed sketches of Jupiter and Mars caught the attention of astronomers at the Lowell Observatory in Flagstaff, Arizona. The observatory, founded by Percival Lowell, was engaged in the search for a ninth planet, known as "Planet X," which Lowell had predicted based on perturbations in the orbits of Uranus and Neptune. In 1929, Tombaugh was invited to join the Lowell

Observatory as a junior astronomer, a position that marked the beginning of his professional career in astronomy.

At the Lowell Observatory, Tombaugh was tasked with conducting a systematic search for Planet X. He used a method known as "blink comparison," which involved taking pairs of photographs of the same section of the night sky several days apart and then rapidly alternating between the two images using a device called a blink comparator. Any object that moved against the relatively static background of stars could be identified as a potential new planet.

Tombaugh's meticulous work paid off on February 18, 1930, when he discovered a moving object on photographic plates taken in January of that year. After confirming the object's movement over subsequent nights, Tombaugh had discovered a new planet. The discovery was announced on March 13, 1930, coinciding with the anniversary of Percival Lowell's birth and the discovery of Uranus by William Herschel. The new planet was named Pluto, after the Roman god of the underworld, a name suggested by an eleven-year-old girl named Venetia Burney from Oxford, England.

The discovery of Pluto catapulted Tombaugh to international fame. He was celebrated as a hero in the scientific community and beyond, receiving numerous awards and honors for his achievement. Despite his newfound fame, Tombaugh remained dedicated to his work and continued his observations at the Lowell Observatory. Over the next decade, he discovered several other celestial objects, including numerous asteroids and variable stars, and conducted extensive surveys of the night sky.

In 1932, Tombaugh married Patricia Edson, and the couple had two children, Annette and Alden. Tombaugh's family life was characterized by a deep bond and mutual support, with Patricia playing a significant role in his career by helping with his research and public engagements.

Tombaugh's discovery of Pluto had a profound impact on the field of astronomy. It expanded our understanding of the solar system and demonstrated the potential for discovering new planets beyond Neptune. Pluto was initially classified as the ninth planet in the solar system, a designation it held for 76 years. However, the discovery of other similar-sized objects in the Kuiper Belt, a region of the solar system beyond Neptune populated by icy bodies, led to a reevaluation of Pluto's status. In 2006, the International Astronomical Union (IAU) reclassified Pluto as a "dwarf planet," a decision that sparked considerable debate and controversy within the astronomical community and among the public.

Tombaugh's contributions to astronomy extended far beyond the discovery of Pluto. After leaving the Lowell Observatory in 1945, he joined the faculty at New Mexico State University (NMSU) in Las Cruces, where he established the university's astronomy program and built a significant research facility. At NMSU, Tombaugh taught and mentored numerous students, inspiring a new generation of astronomers with his dedication and passion for the field. He also continued his research, focusing on the study of minor planets and contributing to the development of observational techniques.

Throughout his career, Tombaugh remained an advocate for space exploration and scientific research. He was involved in the early discussions and planning for space missions to the outer planets, emphasizing the importance of exploring the distant regions of the solar system. His legacy in this regard is evident in the New Horizons mission, which provided the first close-up images of Pluto in 2015, revealing a complex and dynamic world that continues to intrigue scientists.

In addition to his scientific achievements, Tombaugh was known for his humility, integrity, and dedication to education. He believed in the power of science to expand human knowledge and improve society, and he worked tirelessly to share his enthusiasm for astronomy with the public. He gave numerous lectures and public talks, wrote articles for popular science magazines, and engaged with amateur astronomers, always eager to inspire others to look up at the night sky and wonder.

Tombaugh's contributions were recognized with numerous awards and honors throughout his lifetime. He received honorary doctorates from several institutions, including the University of New Mexico and Northern Arizona University. He was a fellow of the American Academy of Arts and Sciences and a member of various astronomical societies. In recognition of his achievements, several astronomical objects have been named in his honor, including the asteroid 1604 Tombaugh and the Tombaugh Regio, a prominent region on Pluto.

Clyde Tombaugh passed away on January 17, 1997, at the age of 90, leaving behind a legacy that continues to inspire astronomers and space enthusiasts around the world. His discovery of Pluto remains one of the most significant milestones in the history of planetary science, and his contributions to the field have had a lasting impact on our understanding of the solar system.

Tombaugh's life story is a testament to the power of curiosity, perseverance, and dedication. From his humble beginnings on a Kansas farm to his groundbreaking discovery and influential career, Tombaugh exemplified the qualities of a true scientist. His relentless pursuit of knowledge, his commitment to education, and his passion for astronomy have left an indelible mark on the field and continue to inspire future generations to explore the cosmos.

Chapter 25: James Van Allen

James Alfred Van Allen, born on September 7, 1914, in Mount Pleasant, Iowa, was a pioneering American space scientist whose contributions to our understanding of Earth's magnetosphere and cosmic radiation profoundly shaped the field of space physics. He is best known for his discovery of the Van Allen radiation belts, a fundamental breakthrough that laid the groundwork for the space age and significantly influenced the design of spacecraft and space missions.

Van Allen's early life was marked by a keen interest in science and exploration. Growing up in a small Midwestern town, he was fascinated by the night sky and the natural world. His parents, who were supportive of his academic pursuits, encouraged his curiosity. Van Allen excelled in his studies and developed a particular interest in physics and astronomy during his high school years.

After graduating from Mount Pleasant High School, Van Allen attended Iowa Wesleyan College, where he earned a bachelor's degree in physics in 1935. He then pursued graduate studies at the University of Iowa, obtaining a master's degree in 1936 and a Ph.D. in 1939. His doctoral research focused on cosmic rays, an interest that would become central to his later work.

Van Allen's career took a significant turn during World War II when he joined the Applied Physics Laboratory (APL) at Johns Hopkins University. At APL, he was involved in the development of the proximity fuse, a crucial innovation for anti-aircraft artillery that greatly enhanced the effectiveness of Allied defenses. His work on the proximity fuse demonstrated his exceptional skills in both experimental physics and engineering, laying the foundation for his later achievements in space science.

After the war, Van Allen returned to the University of Iowa, where he joined the faculty and began to establish himself as a leading researcher in cosmic ray physics. His work during this period included high-altitude balloon experiments designed to study cosmic radiation. These experiments provided valuable data on the intensity and distribution of cosmic rays in the Earth's atmosphere and paved the way for future space-based studies.

Van Allen's interest in high-altitude research led him to collaborate with the U.S. Navy on a series of rocket experiments. Using captured German V-2 rockets, he and his colleagues conducted pioneering studies of the upper atmosphere and cosmic radiation. These experiments marked the beginning of Van Allen's involvement in space research and set the stage for his most significant contributions.

In the early 1950s, Van Allen became involved in the International Geophysical Year (IGY), a global scientific initiative aimed at studying the Earth's geophysical properties. As part of this effort, he proposed a series of high-altitude rocket experiments to investigate the Earth's magnetosphere and cosmic radiation. His proposal was accepted, and he was tasked with developing scientific instruments for these missions.

One of the key innovations developed by Van Allen and his team was the Geiger-Müller tube, a device capable of detecting charged particles in space. These instruments were designed to be launched aboard rockets to study the distribution of cosmic rays and other high-energy particles in the Earth's magnetosphere. Van Allen's team also developed a system for recording and transmitting data from these instruments, a crucial capability for space-based research.

The culmination of Van Allen's efforts came with the launch of the Explorer 1 satellite on January 31, 1958. Explorer 1 was the first successful American satellite, and it carried a suite of scientific instruments designed by Van Allen and his team. The data returned by Explorer 1 revealed the existence of a previously unknown region of intense radiation surrounding the Earth, which became known as the Van Allen radiation belts. This discovery was a landmark achievement in space science, providing the first direct evidence of the complex interactions between the Earth's magnetic field and charged particles from the sun.

The discovery of the Van Allen belts had profound implications for our understanding of the Earth's magnetosphere and the dynamics of space weather. It also highlighted the need to protect spacecraft and astronauts from the potentially harmful effects of radiation. Van Allen's work on the radiation belts earned him widespread recognition and numerous accolades, including the National Medal of Science, the United States' highest scientific honor.

Following the success of Explorer 1, Van Allen continued to play a leading role in space research. He was involved in the design and development of a series of subsequent satellites, including Pioneer 3 and 4, which provided further insights into the radiation belts and other aspects of the space environment. He also served as a key advisor to NASA and other space agencies, helping to shape the direction of space exploration and research.

Throughout his career, Van Allen remained dedicated to education and public outreach. He was a passionate advocate for the importance of scientific research and the need for international collaboration in space exploration. As a professor at the University of Iowa, he mentored numerous students and young researchers, many of whom went on to make significant contributions to the field of space science. His teaching and mentorship helped to establish the University of Iowa as a leading center for space research.

Van Allen's contributions to space science extended beyond his research on the radiation belts. He was also involved in the study of cosmic rays, solar particles, and the interaction between the solar wind and the Earth's magnetosphere. His work helped to lay the foundation for our current understanding of space weather and its impact on Earth and human technology.

In addition to his scientific achievements, Van Allen was known for his integrity, humility, and dedication to the scientific community.

He was a tireless advocate for the importance of basic research and the need to support scientific inquiry. His leadership and vision were instrumental in the establishment of NASA's space science program, and he played a key role in the development of policies and strategies for space exploration.

Van Allen's legacy is reflected in the numerous awards and honors he received throughout his career, including the Gold Medal of the Royal Astronomical Society, the John Adam Fleming Medal of the American Geophysical Union, and the Eddington Medal of the Royal Astronomical Society. He was also elected to the National Academy of Sciences and served as president of the American Geophysical Union.

James Van Allen passed away on August 9, 2006, at the age of 91, leaving behind a legacy of scientific achievement and a lasting impact on the field of space science. His discovery of the Van Allen radiation belts remains one of the most significant milestones in the history of space exploration, and his contributions to our understanding of the Earth's magnetosphere and cosmic radiation continue to shape the field today.

Chapter 26: Roger Penrose

Sir Roger Penrose, born on August 8, 1931, in Colchester, England, is a renowned British mathematician, physicist, and philosopher of science whose groundbreaking work has profoundly influenced our understanding of the universe. Penrose is best known for his contributions to the mathematical foundations of general relativity, cosmology, and the theory of black holes, as well as his work on the nature of consciousness and the fundamental structure of space and time. Over the course of his illustrious career, Penrose has made significant strides in both theoretical physics and the philosophy of science, garnering numerous prestigious awards, including the Nobel Prize in Physics in 2020.

Roger Penrose was born into an intellectually stimulating family. His father, Lionel Penrose, was a geneticist and psychiatrist, while his mother, Margaret Leathes, was a medical doctor. His siblings also pursued academic careers, with his brother Oliver Penrose becoming a distinguished physicist and his brother Jonathan Penrose achieving prominence as a chess grandmaster. This rich intellectual environment undoubtedly influenced Penrose's development and his eventual path in science.

Penrose's formal education began at University College School in London. He then attended University College London, where he initially studied medicine before switching to mathematics, which better suited his interests and talents. He graduated with first-class honors in mathematics in 1952. Following this, Penrose pursued a Ph.D. in algebraic geometry at St John's College, Cambridge, under the supervision of W.V.D. Hodge, completing his doctorate in 1957. His early work focused on the intersection of mathematics and physics, a theme that would dominate his career.

One of Penrose's earliest and most significant contributions came in the form of twistor theory, which he introduced in the 1960s.

Twistor theory is an ambitious mathematical framework aimed at unifying quantum mechanics and general relativity. It represents physical fields, such as those describing gravity and electromagnetism, in a complex space known as twistor space. This approach has provided deep insights into the structure of space-time and has influenced various areas of theoretical physics, though it has yet to achieve its ultimate goal of unification.

In 1965, Penrose published a groundbreaking paper on the nature of black holes, for which he would later receive the Nobel Prize. Building on the work of Albert Einstein and Karl Schwarzschild, Penrose developed the concept of a "trapped surface," a crucial component in the formation of black holes. He demonstrated that once a trapped surface forms, the gravitational collapse is inevitable, leading to a singularity where the known laws of physics break down. This work laid the theoretical foundation for the existence of black holes, providing rigorous mathematical proof of their formation and properties.

Penrose's work on black holes also led to the formulation of the Penrose singularity theorem, which posits that singularities are an inevitable outcome of gravitational collapse in general relativity. This theorem, developed in collaboration with Stephen Hawking, has profound implications for our understanding of the universe. It suggests that the Big Bang, the origin of our universe, might also be a singularity, thus linking the birth of the cosmos to the fundamental laws of physics governing black holes.

Beyond his contributions to black hole theory, Penrose has also made significant advancements in the study of cosmology. He is known for his work on the cosmic censorship conjecture, which posits that singularities resulting from gravitational collapse are hidden within black holes and cannot be observed from the outside. Although this conjecture remains unproven, it has spurred extensive research into the nature of singularities and the fundamental structure of space-time.

In the realm of quantum mechanics, Penrose has proposed several innovative ideas aimed at addressing the fundamental problems of the theory. One of his most notable contributions is the Penrose interpretation of quantum mechanics, which suggests that the collapse of the quantum wave function is a physical process resulting from gravitational effects. This idea, often referred to as objective reduction, posits that gravity plays a crucial role in the transition from quantum superpositions to classical states, thus providing a possible solution to the measurement problem in quantum mechanics.

Penrose's interests extend beyond physics into the philosophy of science and the nature of consciousness. In his book "The Emperor's New Mind" (1989), Penrose argues against the prevailing view that human consciousness can be fully explained by classical computation. He proposes that consciousness involves non-computable processes, possibly related to quantum gravity effects in the brain. This idea, developed further in collaboration with anesthesiologist Stuart Hameroff, led to the controversial Orch-OR (Orchestrated Objective Reduction) theory. Orch-OR posits that microtubules within neurons are sites of quantum processes that contribute to consciousness, challenging conventional computational theories of mind.

In addition to his theoretical work, Penrose has contributed to the field of mathematics through the discovery of Penrose tiling. In the 1970s, he discovered a set of non-periodic tilings that can cover a plane without repeating patterns, now known as Penrose tiles. These tilings have deep implications for the study of quasicrystals and the mathematical understanding of symmetry and structure. Penrose tiles have also captured the public imagination, appearing in various forms of art and architecture.

Penrose's contributions to science and mathematics have been recognized with numerous awards and honors. He was knighted in 1994 for his services to science and has received many prestigious awards, including the Wolf Prize in Physics, the Dirac Medal, the

Copley Medal, and the Albert Einstein Medal. His Nobel Prize in Physics, awarded in 2020, specifically recognized his work on black holes and their implications for our understanding of the universe.

Throughout his career, Penrose has maintained a commitment to public education and outreach. He is a prolific author, having written several popular science books that explain complex scientific concepts to a general audience. These works, including "The Road to Reality" (2004) and "Fashion, Faith, and Fantasy in the New Physics of the Universe" (2016), reflect his deep engagement with the philosophical and foundational issues in physics and cosmology. Penrose's ability to convey intricate scientific ideas in an accessible and engaging manner has made him a prominent figure in the public understanding of science.

Penrose's influence extends beyond his scientific contributions to his role as a mentor and collaborator. He has inspired and worked with many of the leading figures in theoretical physics, including Stephen Hawking, with whom he developed the singularity theorems. His collaborative spirit and willingness to engage with diverse ideas have fostered a rich intellectual environment in which groundbreaking discoveries can flourish.

In recent years, Penrose has continued to explore new frontiers in physics and cosmology. He has proposed the Conformal Cyclic Cosmology (CCC) model, which suggests that the universe undergoes infinite cycles of Big Bangs and subsequent expansions. According to this theory, each cycle, or "aeon," begins with a Big Bang and ends with a conformal transformation that leads to the next Big Bang. This model challenges the conventional view of a singular, one-time Big Bang and offers new insights into the long-term evolution of the universe.

Penrose's CCC model has generated significant interest and debate within the scientific community. It suggests that traces of previous aeons could be detected in the cosmic microwave background radiation, potentially providing observational evidence for the theory.

Although the CCC model remains speculative, it exemplifies Penrose's willingness to explore bold and unconventional ideas in his quest to understand the fundamental nature of the universe.

Throughout his career, Penrose has been characterized by his intellectual curiosity, creativity, and rigorous approach to scientific inquiry. His work bridges the gap between mathematics and physics, providing deep insights into the nature of reality. Penrose's ability to tackle some of the most challenging problems in science, coupled with his philosophical perspective on the implications of his findings, has made him one of the most influential and respected figures in modern theoretical physics.

Chapter 27: Vera Rubin

Vera Cooper Rubin, born on July 23, 1928, in Philadelphia, Pennsylvania, was an American astronomer whose groundbreaking work provided some of the first strong evidence for the existence of dark matter, a mysterious and invisible substance that constitutes most of the mass in the universe. Her pioneering research on the rotation curves of galaxies revolutionized our understanding of the cosmos and significantly influenced the field of astrophysics. Rubin's career was marked by her perseverance in the face of gender barriers and her commitment to fostering the next generation of scientists.

Rubin's interest in astronomy began at a young age. She moved with her family to Washington, D.C., when she was ten years old, where her father, Philip Cooper, an electrical engineer, encouraged her fascination with the night sky. He helped her build a telescope, and she spent countless hours observing the stars and planets. Her early passion for astronomy was further fueled by her reading of science books and her visits to the American Museum of Natural History in New York City.

Despite her enthusiasm for science, Rubin faced significant obstacles as a woman in a male-dominated field. She graduated from high school in 1944 and enrolled at Vassar College, a women's college with a strong tradition in the sciences. At Vassar, she was mentored by the renowned astronomer Maria Mitchell, who had been the first American woman to work as a professional astronomer. Rubin graduated in 1948 with a degree in astronomy and was the only graduate in her class with a degree in that field.

Rubin's ambitions led her to apply to graduate programs in astronomy, but she encountered resistance due to her gender. She initially enrolled at Cornell University, where she studied physics under the guidance of eminent scientists such as Philip Morrison, Richard Feynman, and Hans Bethe. Her master's thesis, completed in 1951, focused on the possibility of bulk motions of galaxies, a controversial topic at the time. Rubin suggested that galaxies might be moving relative to one another, a hypothesis that would later be confirmed and become a key element of modern cosmology.

Rubin's early work did not receive the recognition it deserved, and she faced challenges in securing a position that would allow her to continue her research. Nevertheless, she persisted and earned her Ph.D. in astronomy from Georgetown University in 1954. Her doctoral dissertation, supervised by George Gamow, a prominent cosmologist, focused on the distribution of galaxies and their angular momenta. This work laid the foundation for her later studies on galactic rotation and the dynamics of galaxies.

In the 1960s, Rubin joined the Carnegie Institution's Department of Terrestrial Magnetism, where she began collaborating with Kent Ford, an astronomer known for his expertise in instrumentation. Together, they conducted a series of observations that would lead to one of the most significant discoveries in modern astronomy. Rubin and Ford used a new spectrograph, designed by Ford, to measure the rotation curves of spiral galaxies. A rotation curve plots the velocity of stars and gas in a galaxy as a function of their distance from the galactic center.

Rubin and Ford's observations revealed a surprising and puzzling result. According to Newtonian mechanics and the distribution of visible matter in galaxies, the rotational velocities of stars should decrease with increasing distance from the galactic center, much like the planets in our solar system. However, Rubin and Ford found that the outer regions of galaxies were rotating as fast as or even faster than the inner regions. This discrepancy suggested the presence of an unseen mass, which Rubin hypothesized to be dark matter.

Rubin's findings were initially met with skepticism, but her meticulous work and the accumulation of additional evidence gradually convinced the scientific community. Her research provided some of the first compelling evidence for dark matter, a concept that had been proposed earlier but lacked strong observational support. Dark matter, which does not emit, absorb, or reflect light, interacts gravitationally with visible matter and is believed to make up about 85% of the total mass of the universe.

The implications of Rubin's discovery were profound. It challenged existing models of galactic dynamics and necessitated a reevaluation of the fundamental nature of the universe. Rubin's work on rotation curves became a cornerstone of modern cosmology and influenced subsequent research on dark matter, the large-scale structure of the universe, and the formation and evolution of galaxies.

Despite her significant contributions, Rubin faced ongoing challenges as a woman in science. She encountered discrimination and often felt isolated in a field dominated by men. Nevertheless, she remained committed to her research and to advocating for greater inclusion and diversity in astronomy. Rubin was a strong supporter of women in science and actively mentored young female scientists throughout her career. She emphasized the importance of providing opportunities and support for women and underrepresented groups in STEM fields.

Rubin's achievements were recognized with numerous awards and honors. She was elected to the National Academy of Sciences in 1981 and received the National Medal of Science in 1993, one of the highest honors in American science. She was also awarded the Gold Medal of the Royal Astronomical Society in 1996, becoming the first woman to receive the award since Caroline Herschel in 1828. Despite these accolades, Rubin remained modest and focused on her work, always emphasizing the collaborative nature of scientific discovery.

In addition to her research on galactic rotation, Rubin made significant contributions to other areas of astronomy. She studied the large-scale distribution of galaxies and the structure of the universe, providing insights into the behavior of galaxies in clusters and the influence of dark matter on their dynamics. Her work helped to establish the field of observational cosmology and laid the groundwork for future discoveries about the nature of the universe.

Rubin's legacy extends beyond her scientific contributions. She was a passionate advocate for science education and public outreach, believing that everyone should have the opportunity to engage with and understand the wonders of the universe. She gave numerous public lectures, wrote articles for popular science magazines, and appeared in documentaries, always striving to make astronomy accessible and exciting for the general public.

Rubin's personal life was also marked by her dedication to her family. She married Robert Rubin, a physical chemist, in 1948, and they had four children, all of whom pursued careers in science and medicine. Despite the demands of her career, Rubin maintained a strong commitment to her family and often spoke about the importance of balancing professional and personal responsibilities.

Vera Rubin passed away on December 25, 2016, at the age of 88. Her death was a significant loss to the scientific community, but her legacy lives on through her groundbreaking discoveries and her impact on the field of astronomy. Rubin's work on dark matter fundamentally changed our understanding of the universe and opened new avenues of research that continue to be explored by scientists today.

Chapter 28: Nancy Grace Roman

Nancy Grace Roman was a pioneering American astronomer and a key figure in the field of space science. Often referred to as the "Mother of Hubble," she made significant contributions to NASA and the development of the Hubble Space Telescope. Born on May 16, 1925, in Nashville, Tennessee, Roman showed an early interest in astronomy. Her mother, a music teacher, and her father, a geophysicist, nurtured her curiosity about the natural world. By the time she was 11, Roman had organized an astronomy club among her classmates, demonstrating her early passion for the stars.

Roman pursued her undergraduate degree at Swarthmore College, graduating in 1946 with a degree in astronomy. She then went on to the University of Chicago, where she earned her Ph.D. in 1949. At the University of Chicago, she worked at the Yerkes Observatory under the mentorship of renowned astronomer William W. Morgan. During her time there, she contributed to important research on stellar classification and dynamics, including the study of stars with high velocities and their motions within the galaxy. Her work provided valuable insights into the structure and behavior of our galaxy, the Milky Way.

In 1959, Roman made a career-changing decision to join NASA, which had been established just a year earlier. She became the first Chief of Astronomy in the Office of Space Science, a position she held for nearly two decades. At NASA, Roman was instrumental in planning and advocating for space-based astronomical observatories. She was a visionary who recognized the potential of placing telescopes in space, beyond the distorting effects of Earth's atmosphere. Her efforts were crucial in the early development and eventual success of the Hubble Space Telescope. Roman's work involved extensive collaboration with other scientists, engineers, and policymakers, ensuring that the vision for a space-based observatory could become a reality.

Roman's role at NASA was multifaceted. She was responsible for planning and organizing a series of satellite missions that laid the groundwork for future space telescopes. Her early work included the Orbiting Solar Observatory (OSO) and the Orbiting Astronomical Observatory (OAO) programs. The OAO missions, in particular, provided essential data and experience that informed the design and operation of the Hubble Space Telescope. Roman's leadership and advocacy for the Hubble project were vital during its conceptual phase, as she worked tirelessly to secure funding, support, and technological development necessary for its success.

Beyond her technical and scientific contributions, Roman was a trailblazer for women in science and engineering. At a time when these fields were overwhelmingly dominated by men, Roman broke barriers and paved the way for future generations of women scientists. She was an outspoken advocate for gender equality in the workplace and often spoke about the challenges she faced as a woman in a male-dominated field. Her perseverance and dedication served as an inspiration to many young women aspiring to careers in science and engineering.

Roman's influence extended beyond NASA. She was an active member of various scientific organizations and served on numerous advisory committees and panels. She was a fellow of the American Astronomical Society and the American Association for the Advancement of Science. Her contributions to the field of astronomy were widely recognized, and she received several prestigious awards, including the NASA Exceptional Scientific Achievement Medal in 1969. In her later years, Roman remained active in the scientific community, frequently giving talks and lectures about her work and the importance of space-based astronomy.

Nancy Grace Roman's legacy is indelibly linked with the Hubble Space Telescope, which has revolutionized our understanding of the universe. Launched in 1990, Hubble has provided stunning images and invaluable data, leading to numerous groundbreaking discoveries in astronomy and astrophysics. Roman's vision and efforts in the early days of NASA made this possible, cementing her place as one of the most important figures in the history of space science. She passed away on December 25, 2018, leaving behind a remarkable legacy that continues to inspire and influence the field of astronomy.

Throughout her life, Roman demonstrated a profound commitment to advancing our knowledge of the cosmos. Her work not only expanded our understanding of the universe but also highlighted the importance of perseverance, vision, and advocacy in achieving scientific breakthroughs. Nancy Grace Roman's contributions to space science and her role in shaping the future of astronomy remain an enduring testament to her remarkable career and lasting impact on the field.

Chapter 29: Maria Zuber

Maria T. Zuber is a prominent American geophysicist and planetary scientist whose extensive contributions have significantly advanced our understanding of planetary geophysics and the topography of celestial bodies. Born on June 27, 1958, in Norristown, Pennsylvania, Zuber demonstrated an early aptitude for science and mathematics. She pursued her undergraduate education at the University of Pennsylvania, where she earned a Bachelor of Arts degree in astronomy and geology in 1980. Her early interest in understanding the physical characteristics of planets set the stage for a distinguished career in planetary science.

After completing her undergraduate studies, Zuber continued her academic journey at Brown University, where she received a Master of Science degree in geology in 1983 and a Ph.D. in geophysics in 1986. During her time at Brown, she focused on studying the surface features and internal structures of planets and moons, an area that would become her lifelong passion. Her doctoral research involved the use of remote sensing techniques to analyze the topography and gravity fields of planetary bodies, laying the groundwork for her future contributions to planetary exploration missions.

Zuber began her professional career at Johns Hopkins University's Applied Physics Laboratory and later joined the faculty at the Massachusetts Institute of Technology (MIT) in 1992. At MIT, she quickly established herself as a leading researcher in the field of planetary science. Her work has been characterized by a combination of theoretical modeling and data analysis from space missions, providing critical insights into the geological processes that shape planetary surfaces and interiors.

One of Zuber's most significant contributions has been her involvement in numerous NASA missions that have explored the Moon, Mars, and other celestial bodies. She played a key role in the

Clementine mission, launched in 1994, which mapped the surface of the Moon in unprecedented detail. Zuber's work on Clementine involved analyzing the topographic data to better understand the Moon's surface features and geological history. This mission provided valuable data that informed subsequent lunar exploration efforts.

Zuber's expertise was also crucial to the success of the Mars Global Surveyor (MGS) mission, which launched in 1996. As the principal investigator for the Mars Orbiter Laser Altimeter (MOLA) onboard MGS, she oversaw the collection and analysis of detailed topographic maps of Mars. The MOLA instrument used laser ranging to measure the precise distance between the spacecraft and the Martian surface, creating high-resolution maps that revealed the planet's topography with unprecedented accuracy. These maps have been essential for understanding the geological processes that have shaped Mars, including the role of water and volcanic activity.

Another groundbreaking mission that benefited from Zuber's expertise was the Gravity Recovery and Interior Laboratory (GRAIL) mission, which launched in 2011. GRAIL consisted of two spacecraft that orbited the Moon and measured its gravity field in exquisite detail. As the principal investigator for GRAIL, Zuber led a team of scientists in analyzing the gravity data to infer the Moon's internal structure and composition. The mission provided new insights into the Moon's thermal history, crustal thickness, and the nature of its core, significantly enhancing our understanding of lunar geology.

In addition to her work on specific missions, Zuber has made substantial contributions to the broader field of planetary science through her research on the geophysical processes that shape planetary bodies. She has published numerous influential papers on topics such as the thermal evolution of planets, the formation and dynamics of impact basins, and the tectonic activity on the Moon and Mars. Her research has helped to elucidate the complex interplay between a planet's interior and its surface features, providing a comprehensive understanding of planetary geology.

Zuber's impact extends beyond her scientific contributions; she has also been a trailblazer for women in science and engineering. Throughout her career, she has been a strong advocate for diversity and inclusion in STEM fields, working to create opportunities for underrepresented groups. As the first woman to lead a NASA planetary mission, she has served as a role model and mentor for many young scientists, inspiring them to pursue careers in space science and engineering.

In recognition of her outstanding contributions to planetary science, Zuber has received numerous prestigious awards and honors. She is a member of the National Academy of Sciences, the American Academy of Arts and Sciences, and the American Philosophical Society. She has also been awarded the NASA Distinguished Public Service Medal and the Gerard P. Kuiper Prize from the Division for Planetary Sciences of the American Astronomical Society, among many other accolades.

Beyond her research and advocacy, Zuber has held several influential leadership positions within the scientific community. She has served as the chair of the National Research Council's Space Studies Board and as a member of the President's Council of Advisors on Science and Technology. In 2016, she was appointed as MIT's Vice President for Research, becoming the first woman to hold this position. In this role, she has overseen the institute's research initiatives and has been instrumental in advancing interdisciplinary research and innovation.

Maria Zuber's career is a testament to the power of scientific curiosity, dedication, and leadership. Her contributions to planetary science have fundamentally changed our understanding of the solar system, revealing the intricate details of planetary surfaces and interiors. Her work has not only expanded our knowledge of the planets and moons in our solar system but has also paved the way for future exploration and discovery. As a scientist, mentor, and leader, Zuber continues to inspire and shape the field of planetary science, leaving an indelible mark on the scientific community and the study of our universe.

Chapter 30: Carl Sagan

Carl Sagan was a renowned American astronomer, astrophysicist, cosmologist, author, and science communicator whose work significantly shaped public understanding of science and the universe. Born on November 9, 1934, in Brooklyn, New York, Sagan developed an early fascination with the stars. His interest in astronomy was sparked by his visits to the New York World's Fair and reading about the stars and the universe in science fiction and non-fiction books. His parents supported his curiosity, fostering an environment that encouraged intellectual exploration.

Sagan attended the University of Chicago, where he earned a bachelor's degree in physics in 1954, a master's degree in physics in 1955, and a doctorate in astronomy and astrophysics in 1960. His doctoral thesis, "Physical Studies of Planets," focused on the greenhouse effect on Venus, suggesting that the planet's high surface temperature was due to the trapping of heat by its thick carbon dioxide atmosphere. This work laid the foundation for his lifelong interest in planetary atmospheres and climate change.

After completing his Ph.D., Sagan worked at the Smithsonian Astrophysical Observatory in Cambridge, Massachusetts, and lectured at Harvard University. He also spent time at the University of California, Berkeley, before joining Cornell University in 1968, where he spent the majority of his career. At Cornell, he served as the David Duncan Professor of Astronomy and Space Sciences and was the director of the Laboratory for Planetary Studies.

Sagan's research spanned a wide range of topics within planetary science. He made significant contributions to our understanding of the atmospheres of Venus, Mars, and Jupiter, as well as the seasonal changes on Mars. He was particularly interested in the potential for life elsewhere in the solar system and the universe. His work on the conditions necessary for life and the search for extraterrestrial intelligence (SETI) was groundbreaking. He was one of the first scientists to seriously study the possibility of life on other planets and was instrumental in the development of the field of exobiology.

One of Sagan's most significant scientific contributions was his role in the early space missions to other planets. He was a key advisor to NASA on the Mariner, Viking, Voyager, and Galileo missions. He played a crucial role in designing experiments and interpreting data from these missions. For example, he worked on the imaging team for the Viking missions to Mars in the 1970s, which conducted experiments searching for signs of life on the Martian surface. Although the results were inconclusive, the missions provided a wealth of information about Mars' geology and climate.

Sagan was also deeply involved in the Voyager missions, which launched in 1977 to explore the outer planets. He helped select the iconic "Golden Record" carried by the Voyager spacecraft, which contained sounds and images intended to represent the diversity of life and culture on Earth to any extraterrestrial intelligence that might find it. The record included greetings in 55 languages, music from various cultures and eras, and sounds of nature. Sagan's work on the Golden Record reflected his belief in the importance of communication and understanding across the cosmos.

In addition to his scientific research, Carl Sagan was a prolific author and one of the most effective science communicators of the 20th century. He wrote more than 20 books, both popular science and science fiction. His 1980 book "Cosmos," which accompanied the television series of the same name, became one of the best-selling science books ever published. "Cosmos" explored a wide range of scientific topics, from the origins of life and the solar system to the search for extraterrestrial life and the future of humanity. The television series, also hosted by Sagan, reached hundreds of millions of viewers worldwide and won numerous awards. It played a crucial role in inspiring a new generation of scientists and enthusiasts.

Sagan's ability to communicate complex scientific ideas in an accessible and engaging manner made him a beloved public figure. He was a frequent guest on television shows and radio programs, where he discussed topics ranging from space exploration to climate change and the importance of scientific thinking. His phrase "billions and billions," though he never actually used it in his work, became a popular catchphrase associated with him, highlighting the vastness of the universe.

Sagan's novel "Contact," published in 1985, is another testament to his impact on both science and popular culture. The story, which explores the themes of communication with extraterrestrial intelligence and the interplay between science and religion, was later adapted into a successful film starring Jodie Foster. "Contact" reflected Sagan's lifelong fascination with the possibility of life beyond Earth and the philosophical implications of such a discovery.

Throughout his career, Sagan was an advocate for scientific skepticism and critical thinking. He was a co-founder of the Planetary Society in 1980, an organization dedicated to the exploration of the solar system and the search for extraterrestrial life. The society aimed to engage the public in space exploration and to advocate for robust funding for space missions. Sagan's commitment to public engagement and advocacy for science policy was evident in his efforts to educate policymakers and the public about the importance of space exploration and environmental protection.

One of Sagan's most profound contributions to the public understanding of science was his emphasis on the fragility and uniqueness of Earth. His reflections on the famous "Pale Blue Dot" photograph, taken by the Voyager 1 spacecraft as it looked back at Earth from the edge of the solar system, emphasized the need for humanity to cherish and protect our planet. In his 1994 book "Pale Blue Dot: A Vision of the Human Future in Space," Sagan eloquently articulated the idea that the Earth is a tiny, fragile oasis in the vast expanse of the cosmos, and that our survival depends on our ability to live sustainably and peacefully.

Sagan's concern for the environment extended to his warnings about the dangers of nuclear war and climate change. He was a vocal advocate for nuclear disarmament and worked to raise awareness about the potential consequences of nuclear conflict. His collaboration with other scientists on the concept of "nuclear winter" demonstrated the potential for catastrophic climate effects resulting from nuclear war, influencing public and policy discussions on the subject.

Despite his immense contributions to science and public understanding, Sagan faced criticism and controversy. Some of his peers criticized him for his outspoken advocacy and media presence, arguing that he sometimes oversimplified complex scientific issues. However, his ability to engage and inspire the public was undeniable, and his legacy as a science communicator remains influential.

Carl Sagan's life and work left an indelible mark on the fields of astronomy and planetary science, as well as on popular culture. His efforts to promote scientific literacy, critical thinking, and the exploration of the cosmos inspired millions of people around the world. He received numerous awards and honors throughout his career, including the Pulitzer Prize for his book "The Dragons of Eden," the National Academy of Sciences Public Welfare Medal, and the NASA Distinguished Public Service Medal. Sagan passed away on December 20, 1996, but his vision and legacy continue to inspire future generations of scientists, educators, and space enthusiasts.

Chapter 31: Edward Witten

Edward Witten is a preeminent theoretical physicist whose work has significantly advanced the fields of mathematical physics, string theory, and quantum field theory. Born on August 26, 1951, in Baltimore, Maryland, Witten displayed an early aptitude for mathematics and science. His father, Louis Witten, was a theoretical physicist, which undoubtedly influenced his intellectual environment. Witten's academic journey is remarkable not only for his profound contributions to theoretical physics but also for his unconventional path through higher education.

Witten began his undergraduate studies at Brandeis University, initially majoring in history with a minor in linguistics. However, his interest in mathematics and physics gradually overshadowed his initial academic pursuits. He completed his bachelor's degree in history in 1971, but his fascination with the fundamental laws of nature led him to switch fields. He pursued graduate studies in applied mathematics at Princeton University, where he earned a Ph.D. in 1976 under the supervision of David Gross, a future Nobel laureate in physics.

Witten's early work focused on aspects of quantum field theory, particularly gauge theory and supersymmetry. In the late 1970s and early 1980s, he made groundbreaking contributions to the understanding of the relationship between gauge theories and the topology of four-dimensional manifolds. His research demonstrated how non-abelian gauge theories could be used to uncover deep mathematical structures, which would later influence the development of topological quantum field theory.

One of Witten's most significant early contributions was his work on the Atiyah-Singer index theorem, which connects quantum field theory with differential geometry. The index theorem provides a method for counting the solutions to certain differential equations and has profound implications in both mathematics and theoretical physics. Witten's insights into the physical interpretation of this theorem helped bridge the gap between the two disciplines and opened new avenues of research.

Witten's reputation as a leading theoretical physicist was firmly established in the 1980s with his contributions to string theory, a framework that aims to unify all fundamental forces of nature, including gravity, by modeling particles as one-dimensional objects called strings. String theory had been developed in the late 1960s and early 1970s, but it faced numerous challenges and inconsistencies. Witten's work provided crucial insights that helped address these issues and significantly advanced the field.

In 1984, Witten, along with Michael Green and John Schwarz, played a pivotal role in the "first superstring revolution" by demonstrating the anomaly cancellation in type I string theory. This breakthrough showed that string theory could be a consistent theory of quantum gravity, rekindling interest in the field. Witten's work on string theory also led to the discovery of new mathematical structures, such as mirror symmetry and duality, which have profound implications for both physics and mathematics.

Mirror symmetry, for example, is a phenomenon in string theory where two different Calabi-Yau manifolds can give rise to equivalent physical theories. This unexpected relationship between seemingly different geometrical spaces has led to significant advances in algebraic geometry and has provided new tools for solving complex mathematical problems. Witten's contributions to understanding these symmetries have been instrumental in bridging the gap between theoretical physics and pure mathematics.

In 1990, Witten became the first physicist to be awarded the Fields Medal, the most prestigious award in mathematics, for his work on the interplay between geometry and quantum field theory. The Fields Medal is typically awarded to mathematicians under the age of 40, and Witten's selection underscored his profound impact on both fields. His receipt of the Fields Medal highlighted the deep connections between mathematics and theoretical physics and brought greater recognition to the importance of interdisciplinary research.

Witten's influence extended further during the "second superstring revolution" in the mid-1990s, which was marked by the discovery of dualities between different string theories. These dualities suggested that the five previously known consistent string theories were actually different manifestations of a single underlying theory, known as M-theory. Witten's work was pivotal in elucidating these dualities and in proposing that M-theory, which incorporates 11-dimensional supergravity, could be a unifying framework for all fundamental forces and particles.

In addition to his contributions to string theory, Witten has made significant advances in quantum field theory and topological field theory. His work on topological quantum field theory has provided new insights into the classification of different phases of matter and has had far-reaching implications in condensed matter physics. Witten's formulation of Chern-Simons theory, a topological quantum field theory, has been particularly influential in understanding the quantum Hall effect and other phenomena in low-dimensional systems.

Witten's ability to bridge the gap between mathematics and physics is perhaps best exemplified by his development of the Seiberg-Witten theory in the mid-1990s. This theory, which builds on earlier work by Nathan Seiberg, provides a powerful framework for understanding the non-perturbative aspects of supersymmetric gauge theories. The Seiberg-Witten equations have deep mathematical significance and have led to important results in four-dimensional topology, particularly in the study of smooth structures on four-manifolds.

Throughout his career, Witten has been known for his exceptional mathematical intuition and his ability to tackle some of the most challenging problems in theoretical physics. His work has been characterized by a deep understanding of both the physical principles and the mathematical structures underlying the theories he studies. This unique combination of skills has enabled him to make groundbreaking contributions that have reshaped our understanding of the fundamental nature of the universe.

Witten's influence extends beyond his research contributions; he has also been a dedicated educator and mentor. As a professor at the Institute for Advanced Study in Princeton, New Jersey, Witten has trained and inspired numerous students and postdoctoral researchers who have gone on to make significant contributions to theoretical physics and mathematics. His lectures and writings have been widely regarded for their clarity and depth, helping to disseminate complex ideas to a broader audience.

Witten has received numerous awards and honors in recognition of his contributions to science. In addition to the Fields Medal, he has been awarded the MacArthur Fellowship, also known as the "genius grant," the Dirac Medal, the Albert Einstein Medal, and the Breakthrough Prize in Fundamental Physics, among many others. He is a member of several prestigious scientific organizations, including the National Academy of Sciences, the American Academy of Arts and Sciences, and the Royal Society.

Despite his many accolades, Witten has remained modest and focused on his work. He is known for his collaborative spirit and his willingness to engage with colleagues across disciplines. His contributions have not only advanced our understanding of the fundamental laws of nature but have also fostered greater collaboration between mathematicians and physicists, enriching both fields.

Witten's work continues to influence contemporary research in theoretical physics and mathematics. The concepts and techniques he has developed are integral to ongoing efforts to understand the unification of fundamental forces, the nature of space-time, and the origins of the universe. His insights into string theory, quantum field theory, and mathematical physics have laid the groundwork for future discoveries and have inspired a generation of scientists to pursue these challenging and exciting questions.

Chapter 32: Katherine Johnson

Katherine Johnson was an extraordinary African American mathematician whose work at NASA was pivotal in the success of numerous space missions, including the historic Apollo moon landings. Born on August 26, 1918, in White Sulphur Springs, West Virginia, Johnson displayed remarkable mathematical abilities from a young age. Her father, Joshua Coleman, was a farmer and a handyman, and her mother, Joylette Coleman, was a teacher. Recognizing Katherine's prodigious talent, her parents ensured she received the best education possible, even moving the family to Institute, West Virginia, so she could attend high school at West Virginia State College, an all-black institution, because educational opportunities for African Americans were limited in her hometown.

Katherine Johnson graduated from high school at the age of 14 and entered West Virginia State College. There, she was mentored by the prominent African American mathematician Dr. William W. Schieffelin Claytor, who created a special course in analytic geometry just for her. Johnson graduated summa cum laude in 1937, with degrees in mathematics and French, and began her career as a schoolteacher. In 1939, she was selected as one of the first three African American students to attend graduate school at West Virginia University. However, she left the program to start a family with her husband, James Goble.

In 1953, Johnson began working at the National Advisory Committee for Aeronautics (NACA), the precursor to NASA, at the Langley Research Center in Hampton, Virginia. She joined the West Area Computing unit, a group of African American women mathematicians who manually performed complex calculations. These women were known as "computers." Johnson's talent and expertise soon set her apart. In a time when racial segregation and gender discrimination were rampant, Johnson's determination and skill earned her a place on the Flight Research Division's team, where she analyzed data from flight tests.

Johnson's first major assignment at NACA was to calculate the trajectory for Alan Shepard's Freedom 7 mission in 1961, America's first human spaceflight. Her calculations were crucial to ensuring that Shepard's spacecraft would follow a precise trajectory, allowing for a safe re-entry and recovery. This mission marked a significant achievement for the United States in the Space Race against the Soviet Union.

Her expertise was further highlighted during John Glenn's orbital flight in 1962. Glenn, the first American to orbit the Earth, specifically requested that Johnson verify the calculations made by the electronic computers. "Get the girl to check the numbers," he famously said. Johnson meticulously checked the orbital equations and confirmed the accuracy of the computer's results, ensuring the mission's success. Glenn's flight was a turning point in the space race and cemented Johnson's reputation as a brilliant mathematician.

Johnson's contributions to the Apollo program were even more significant. She was responsible for calculating the trajectories for the Apollo 11 mission, which landed the first humans on the moon in 1969. Her work ensured that the Lunar Module, "Eagle," could successfully land on the moon and return to the Command Module, "Columbia," for the trip back to Earth. Johnson also worked on the emergency return trajectory for Apollo 13, a mission that encountered significant challenges after an oxygen tank explosion. Her calculations helped devise a safe path for the astronauts' return, demonstrating her crucial role in the mission's eventual success.

In addition to her work on crewed spaceflights, Johnson contributed to the development of the Space Shuttle program and the Earth Resources Satellite, which collected data on the Earth's natural resources and environment. Her work extended beyond calculations, as she also co-authored 26 scientific papers, contributing to the broader body of knowledge in the field of space science and engineering. Johnson's publications, which often focused on the mathematical underpinnings of spacecraft trajectories and orbital mechanics, are still referenced by researchers and engineers today.

Johnson's achievements are particularly remarkable given the social and institutional barriers she faced as an African American woman in a predominantly white, male-dominated field. She broke through these barriers with her exceptional talent, perseverance, and dedication to her work. Johnson often credited her success to her parents' emphasis on education and her own relentless curiosity and love for mathematics. She also cited the importance of mentors and supportive colleagues who recognized her abilities and provided opportunities for her to excel.

Katherine Johnson's legacy extends beyond her technical contributions to NASA's missions. Her story has become a powerful symbol of the importance of diversity and inclusion in science, technology, engineering, and mathematics (STEM) fields. She inspired countless individuals, particularly women and minorities, to pursue careers in these areas. Johnson received numerous awards and honors in recognition of her contributions, including the Presidential Medal of Freedom, awarded by President Barack Obama in 2015. This award, the highest civilian honor in the United States, acknowledged her exceptional contributions to the nation's space program and her role as a trailblazer for women and minorities in STEM.

In 2016, Johnson's story was brought to a wider audience through the book "Hidden Figures" by Margot Lee Shetterly, which was later adapted into an acclaimed film of the same name. The book and film highlighted the contributions of Johnson and her colleagues Dorothy Vaughan and Mary Jackson, who also played crucial roles at NASA. "Hidden Figures" shed light on the often-overlooked contributions of African American women to the space program and underscored the importance of recognizing diverse talents in scientific achievements.

Johnson continued to be an advocate for STEM education throughout her life, encouraging young people to pursue their interests in mathematics and science. She spoke at schools, universities, and public events, sharing her story and emphasizing the importance of hard work, perseverance, and the pursuit of knowledge. Her impact on education and public awareness of the contributions of women and minorities to science and technology cannot be overstated.

In addition to the Presidential Medal of Freedom, Johnson received numerous other accolades. She was inducted into the National Women's Hall of Fame and the West Virginia Education Hall of Fame. In 2019, NASA named a new computational research facility at the Langley Research Center in her honor, further cementing her legacy within the organization. The Katherine G. Johnson Computational Research Facility serves as a testament to her contributions and a reminder of the vital role of diversity in innovation and scientific progress.

Katherine Johnson passed away on February 24, 2020, at the age of 101. Her life and work left an indelible mark on the history of space exploration and the broader scientific community. Her contributions to the success of America's space missions, her role as a pioneer for women and minorities in STEM, and her dedication to education and public outreach continue to inspire future generations. Johnson's story is a powerful reminder of the impact that one individual can have on advancing human knowledge and breaking down barriers, paving the way for a more inclusive and innovative future.

Chapter 33: Stephen Hawking

Stephen Hawking was one of the most renowned theoretical physicists and cosmologists of the 20th and 21st centuries. His work on black holes, the nature of the universe, and the fundamental laws of physics has had a profound impact on our understanding of the cosmos. Hawking was not only a brilliant scientist but also a best-selling author and a prominent public figure who brought complex scientific ideas to a broader audience. His life story, marked by his battle with a debilitating illness, is a testament to human resilience and intellectual achievement.

Stephen William Hawking was born on January 8, 1942, in Oxford, England. His birth coincided with the 300th anniversary of the death of Galileo Galilei, an event Hawking often remarked upon with a sense of destiny. He grew up in St. Albans, a town north of London, where he demonstrated an early interest in science and the stars. Hawking's father, Frank Hawking, was a research biologist, and his mother, Isobel Hawking, was a politically active woman who encouraged intellectual pursuits.

Hawking's academic journey began at St. Albans School, where he was a good, though not exceptional, student. His peers and teachers recognized his keen intellect, and he was nicknamed "Einstein" due to his interest in science and mathematics. In 1959, he enrolled at University College, Oxford, his father's alma mater, to study physics. Although he found the coursework relatively easy and did not initially put in much effort, he graduated with a first-class honors degree in natural sciences.

In 1962, Hawking began his graduate work at Trinity Hall, Cambridge, where he studied cosmology under the supervision of Dennis Sciama, one of the leading figures in the field. It was during his time at Cambridge that Hawking's health began to deteriorate. In 1963, he was diagnosed with amyotrophic lateral sclerosis (ALS), also known as Lou Gehrig's disease, a rare and progressive neurodegenerative condition that affects nerve cells in the brain and spinal cord. Doctors gave him a prognosis of only two years to live.

Despite this devastating diagnosis, Hawking continued his research with renewed determination. His condition progressed more slowly than initially expected, allowing him to make significant contributions to theoretical physics. In 1966, he completed his doctoral thesis on the properties of expanding universes, which earned him a Ph.D. and recognition as a rising star in the field of cosmology.

One of Hawking's earliest and most notable achievements was his work on singularity theorems in collaboration with mathematician Roger Penrose. Using the principles of general relativity, they demonstrated that singularities—regions of space-time where gravitational forces cause matter to have infinite density and zero volume—are a general feature of all solutions to Einstein's equations of general relativity that describe black holes. This work showed that the universe itself must have begun as a singularity, a point of infinite density, at the moment of the Big Bang. Their findings were groundbreaking and provided strong support for the Big Bang theory of cosmology.

In the early 1970s, Hawking turned his attention to black holes, regions of space-time where gravity is so strong that nothing, not even light, can escape. His most famous contribution to the field came in 1974 when he proposed that black holes are not entirely black but emit radiation due to quantum effects near the event horizon, the boundary beyond which nothing can escape. This theoretical prediction, known as Hawking radiation, was revolutionary because it combined principles from quantum mechanics, general relativity, and thermodynamics. Hawking's work suggested that black holes could eventually evaporate and disappear, challenging the prevailing understanding of their permanence.

Hawking's radiation provided a potential mechanism for black hole entropy, aligning with the second law of thermodynamics, which states that the total entropy, or disorder, of a closed system can never decrease. This insight led to the formulation of the laws of black hole thermodynamics and has had profound implications for our understanding of the fundamental nature of the universe.

In 1979, Hawking was appointed the Lucasian Professor of Mathematics at the University of Cambridge, a prestigious position once held by Isaac Newton. This appointment recognized his outstanding contributions to theoretical physics and cosmology. Despite his worsening physical condition, which eventually left him almost completely paralyzed and dependent on a computerized speech synthesizer for communication, Hawking continued to produce groundbreaking work and collaborate with other scientists.

Hawking's impact extended beyond the academic world. In 1988, he published "A Brief History of Time: From the Big Bang to Black Holes," a popular science book that explained complex concepts in cosmology in an accessible manner. The book became an international bestseller, selling over 10 million copies and remaining on the Sunday Times bestseller list for a record-breaking 237 weeks. "A Brief History of Time" brought Hawking global fame and made him a household name. His ability to convey profound scientific ideas with clarity and wit captivated readers worldwide and sparked widespread interest in the nature of the universe.

Hawking followed up with several other books aimed at a general audience, including "The Universe in a Nutshell" (2001), "A Briefer History of Time" (2005), co-authored with Leonard Mlodinow, and "The Grand Design" (2010), also co-authored with Mlodinow. In "The Grand Design," Hawking argued that the laws of physics alone could explain the creation of the universe, a viewpoint that stirred considerable debate and discussion.

Throughout his career, Hawking continued to explore the fundamental questions of cosmology and theoretical physics. He investigated the nature of black holes, the origins of the universe, and the potential for a unified theory of physics that could reconcile general relativity with quantum mechanics, often referred to as the "Theory of Everything." One of his later contributions involved the concept of "cosmic inflation," the idea that the universe underwent a rapid expansion shortly after the Big Bang, which helped explain the large-scale structure of the cosmos.

Hawking's work on the information paradox, related to what happens to information that falls into a black hole, remained a central theme of his research. This paradox arises from the apparent conflict between quantum mechanics, which suggests that information must be preserved, and general relativity, which implies that information could be lost in a black hole. Hawking's initial belief that information could be lost in black holes was later revised, and he proposed that information might be preserved in some form, a topic that continues to be an area of active research and debate among physicists.

Despite his severe physical limitations, Hawking remained an active participant in the scientific community, giving lectures, attending conferences, and engaging with fellow researchers. He also made several public appearances and cameo appearances in popular media, including television shows like "Star Trek: The Next Generation," "The Simpsons," and "The Big Bang Theory." These appearances helped cement his status as a cultural icon and popularizer of science.

Hawking's personal life was also marked by notable events. He married Jane Wilde in 1965, and they had three children: Robert, Lucy, and Timothy. Jane played a significant role in Hawking's life, providing support and care during the early years of his illness. Their marriage, however, faced numerous challenges due to the strains of his illness and the demands of his career. They divorced in 1995. Shortly after,

Hawking married Elaine Mason, one of his nurses. This marriage ended in 2006 amidst allegations of abuse, which Hawking denied.

In his later years, Hawking continued to advocate for scientific research and public engagement with science. He spoke on issues such as the potential risks posed by artificial intelligence, the future of humanity in space, and the importance of addressing climate change. He warned of the existential threats humanity might face and emphasized the need for collective action to ensure the survival and prosperity of our species.

Stephen Hawking passed away on March 14, 2018, at the age of 76. His death marked the loss of one of the most brilliant minds of our time, but his legacy lives on through his contributions to science and his impact on popular culture. Hawking's work has inspired countless scientists and thinkers, and his books continue to educate and inspire people around the world about the wonders of the universe.

Hawking's life story is not just one of intellectual achievement but also one of extraordinary resilience and determination. Despite being diagnosed with a debilitating illness at a young age and given a grim prognosis, he went on to make groundbreaking contributions to our understanding of the cosmos. His ability to overcome immense physical challenges and continue his scientific work is a testament to his remarkable spirit and dedication.

In recognition of his contributions, Hawking received numerous awards and honors throughout his life, including the Copley Medal from the Royal Society, the Presidential Medal of Freedom from the United States, and the Fundamental Physics Prize. He was a Fellow of the Royal Society and a member of the US National Academy of Sciences, among many other accolades.

Chapter 34: Neil Armstrong

Neil Armstrong, the first human to set foot on the moon, is one of the most iconic figures in the history of space exploration. His life story is a tale of extraordinary achievement, relentless dedication, and remarkable courage. Armstrong's contributions to aviation and space travel extend beyond his historic moonwalk, encompassing a career marked by numerous pioneering accomplishments.

Neil Alden Armstrong was born on August 5, 1930, in Wapakoneta, Ohio. His fascination with flight began early, fueled by a ride in a Ford Trimotor airplane when he was just six years old. This early experience sparked a lifelong passion for aviation. By the age of 16, Armstrong had already earned his student pilot's license, even before he received his driver's license. His commitment to flying was evident from a young age, and it would define his career and legacy.

Armstrong enrolled at Purdue University in 1947, pursuing a degree in aeronautical engineering. His education was interrupted by the Korean War, during which he served as a naval aviator. Armstrong flew 78 combat missions, demonstrating exceptional skill and bravery. He returned to Purdue in 1952 to complete his degree, graduating in 1955. His wartime service honed his piloting abilities and underscored his dedication to aviation.

After graduation, Armstrong joined the National Advisory Committee for Aeronautics (NACA), which later became NASA. He worked as a test pilot at the High-Speed Flight Station at Edwards Air Force Base in California. Armstrong's work involved flying experimental aircraft, including the rocket-powered X-15, which could reach speeds of up to 4,000 miles per hour and altitudes beyond 200,000 feet. This period was critical in developing Armstrong's expertise in high-speed and high-altitude flight, skills that would prove invaluable in his later career as an astronaut.

In 1962, Armstrong was selected as one of NASA's second group of astronauts, known as the "New Nine." This group was integral to America's efforts to land a man on the moon. Armstrong's first spaceflight was as command pilot of Gemini 8 in March 1966. The mission was to perform the first docking of two spacecraft in orbit, a crucial step for the planned lunar missions. During the mission, Armstrong and his fellow astronaut, David Scott, successfully docked with an unmanned Agena target vehicle. However, the mission encountered a critical problem when a stuck thruster caused the combined spacecraft to spin uncontrollably. Armstrong's quick thinking and calm demeanor allowed him to regain control of the spacecraft, abort the mission, and safely return to Earth. This incident showcased Armstrong's exceptional piloting skills and his ability to remain composed under pressure.

Armstrong's next mission would cement his place in history. He was selected as the commander of Apollo 11, the first manned mission to land on the moon. On July 16, 1969, Armstrong, along with lunar module pilot Edwin "Buzz" Aldrin and command module pilot Michael Collins, launched from Kennedy Space Center atop a Saturn V rocket. Four days later, on July 20, Armstrong and Aldrin descended to the lunar surface in the Lunar Module, "Eagle," while Collins remained in orbit around the moon.

The landing itself was fraught with challenges. As the Lunar Module approached the surface, Armstrong realized that the pre-programmed landing site was strewn with large boulders. Taking manual control, he carefully piloted the Eagle to a safer location, with only seconds of fuel remaining. His skillful landing demonstrated his remarkable piloting abilities and composure.

At 10:56 p.m. EDT on July 20, 1969, Armstrong descended the ladder of the Lunar Module and set foot on the moon, uttering the famous words, "That's one small step for man, one giant leap for mankind." This moment, broadcast live to a global audience, marked a significant milestone in human history. Aldrin joined him shortly after, and the two astronauts spent about two and a half hours exploring the lunar surface, collecting samples, and conducting experiments.

The success of Apollo 11 was a culmination of years of planning, research, and training. It represented the achievement of President John F. Kennedy's goal, set in 1961, of landing a man on the moon and returning him safely to Earth before the end of the decade. The mission was not only a triumph of technology and engineering but also a symbol of human ingenuity and the relentless pursuit of knowledge.

Upon returning to Earth, Armstrong and his crewmates were celebrated as heroes. They were honored with parades, awards, and global recognition. Armstrong received the Presidential Medal of Freedom, the highest civilian award in the United States, among numerous other accolades. Despite the fame, Armstrong remained humble and private, often attributing the success of the mission to the collective efforts of thousands of scientists, engineers, and support staff.

After Apollo 11, Armstrong took on various roles within NASA and academia. He served as Deputy Associate Administrator for Aeronautics at NASA Headquarters in Washington, D.C., before resigning from NASA in 1971. He then became a professor of aerospace engineering at the University of Cincinnati, a position he held until 1979. Armstrong's teaching career allowed him to share his knowledge and experience with the next generation of engineers and pilots.

Armstrong's post-NASA years were marked by a commitment to education, exploration, and public service. He served on several corporate boards and participated in various commissions, including the Rogers Commission, which investigated the Space Shuttle Challenger disaster in 1986. Armstrong's insights and expertise were invaluable in understanding the causes of the tragedy and recommending safety improvements for future missions.

Despite his public role, Armstrong was known for his private nature. He rarely sought the spotlight and preferred to focus on his work and family. His humility and modesty were consistent with his character, and he often downplayed his achievements, emphasizing the collaborative nature of the space program.

In addition to his professional accomplishments, Armstrong had a rich personal life. He married Janet Shearon in 1956, and they had three children: Eric, Karen, and Mark. Tragically, Karen died of a brain tumor in 1962, a loss that deeply affected Armstrong and his family. Armstrong and Janet divorced in 1994, and he later married Carol Knight in 1994. Armstrong enjoyed spending time with his family, flying planes, and engaging in various hobbies, including farming and flying gliders.

Neil Armstrong's legacy is profound and far-reaching. His historic moonwalk remains a defining moment in human history, symbolizing the limitless potential of human exploration and achievement. Armstrong's contributions to aviation and space travel have inspired generations of scientists, engineers, and explorers. His life story, marked by courage, dedication, and humility, continues to resonate with people around the world.

In recognition of his contributions, numerous institutions and landmarks have been named in his honor. These include schools, museums, and airports, reflecting the widespread admiration and respect for his achievements. The lunar crater "Armstrong" and the asteroid "6469 Armstrong" are named after him, ensuring his legacy extends beyond Earth.

Armstrong passed away on August 25, 2012, at the age of 82, following complications from heart surgery. His death was met with an outpouring of tributes from around the world, highlighting his impact on science, exploration, and human achievement. President Barack Obama described Armstrong as "among the greatest of American heroes—not just of his time, but of all time."

In the years since his passing, Armstrong's legacy has continued to inspire. His story is a reminder of what can be achieved through perseverance, dedication, and a relentless pursuit of knowledge. As humanity looks to the future of space exploration, including potential missions to Mars and beyond, Armstrong's pioneering spirit serves as a guiding light.

Armstrong's impact on popular culture is also significant. His moon landing has been the subject of numerous books, documentaries, and films, including the acclaimed movie "First Man" (2018), which chronicles his life and the Apollo 11 mission. These portrayals have helped to keep his story alive for new generations and underscore the enduring significance of his achievements.

Chapter 35: Buzz Aldrin

Buzz Aldrin, born Edwin Eugene Aldrin Jr. on January 20, 1930, in Montclair, New Jersey, is an iconic figure in the history of space exploration, best known for being one of the first two humans to walk on the moon during the Apollo 11 mission in 1969. Aldrin's life story is a remarkable tale of achievement, dedication, and perseverance, encompassing his early years, military career, groundbreaking contributions to space exploration, and his ongoing advocacy for human space travel.

Aldrin's fascination with aviation and space began at an early age, heavily influenced by his father, Edwin Eugene Aldrin Sr., who was a military aviator and a student of rocket pioneer Robert Goddard. This early exposure to flight and technology set the stage for Aldrin's future career. After graduating from Montclair High School, Aldrin attended the United States Military Academy at West Point, where he excelled academically and athletically, graduating third in his class in 1951 with a degree in mechanical engineering.

Following his graduation from West Point, Aldrin joined the United States Air Force. He flew 66 combat missions during the Korean War, piloting F-86 Sabre jets and shooting down two MiG-15 aircraft. His performance in combat earned him the Distinguished Flying Cross for his bravery and skill. After the war, Aldrin continued his career as a pilot, becoming an instructor and then a test pilot. His exceptional flying skills and technical knowledge made him a standout in the Air Force.

In 1959, Aldrin earned a Doctorate of Science in Astronautics from the Massachusetts Institute of Technology (MIT). His doctoral thesis focused on guidance for manned orbital rendezvous, a critical concept for the future of space travel, especially for the Apollo missions that required precise docking procedures in space. Aldrin's academic work earned him the nickname "Dr. Rendezvous," and his expertise in this area would later prove invaluable to NASA's space program.

In 1963, Aldrin was selected as one of the third group of astronauts by NASA. His first spaceflight was as pilot of Gemini 12 in November 1966, alongside command pilot Jim Lovell. The mission's primary objectives included docking with an Agena target vehicle and performing extravehicular activities (EVAs), or spacewalks. Aldrin's innovative techniques for EVA, including the use of handrails and footholds, were instrumental in overcoming the challenges of working in the vacuum of space. He spent a total of 5 hours and 30 minutes outside the spacecraft, setting a new record and demonstrating the feasibility of working on the exterior of a spacecraft, a necessary capability for the Apollo missions.

Aldrin's most famous mission came in July 1969 with Apollo 11. As the Lunar Module pilot, Aldrin, along with mission commander Neil Armstrong and command module pilot Michael Collins, embarked on the historic journey to the moon. On July 20, 1969, Armstrong and Aldrin descended to the lunar surface in the Lunar Module "Eagle." After Armstrong's famous first step, Aldrin followed, becoming the second human to walk on the moon. Aldrin described the lunar surface as "magnificent desolation," a phrase that captured both the beauty and the barren, untouched nature of the moon.

During their time on the lunar surface, Armstrong and Aldrin conducted scientific experiments, collected rock and soil samples, and took numerous photographs. They planted the American flag and left behind a plaque that read, "Here men from the planet Earth first set foot upon the Moon, July 1969 A.D. We came in peace for all mankind." The successful landing and return of Apollo 11 marked a monumental achievement in human history, fulfilling President John F. Kennedy's goal of landing a man on the moon and returning him safely to Earth before the end of the 1960s.

The impact of Apollo 11 extended beyond the scientific and technological achievements; it was a unifying moment for humanity. Millions of people around the world watched the moon landing live on television, inspired by the extraordinary feat. Aldrin, Armstrong, and Collins were celebrated as heroes, receiving accolades and honors from around the globe. Aldrin's contributions to the mission, particularly his expertise in orbital mechanics and rendezvous techniques, were crucial to its success.

After the Apollo 11 mission, Aldrin continued to serve NASA in various capacities, but he struggled with the transition from the intense focus of space missions to the more mundane aspects of life on Earth. In 1971, he retired from NASA and the Air Force, where he had achieved the rank of Colonel. His post-NASA years were marked by personal challenges, including struggles with depression and alcoholism. Aldrin has been open about these difficulties, using his experiences to advocate for mental health awareness and support for those facing similar issues.

Despite these personal challenges, Aldrin remained an influential figure in the space community. He authored several books, including his autobiography "Return to Earth" (1973) and "Magnificent Desolation" (2009), as well as science fiction novels and works on space exploration. Aldrin's writings often reflected his unwavering belief in the importance of human space exploration and his vision for the future.

Aldrin has been a tireless advocate for the human exploration of Mars. He developed the "Aldrin Mars Cycler," a spacecraft system designed to facilitate regular travel between Earth and Mars. His advocacy efforts include public speaking, consulting, and collaborating with space agencies and private companies to promote missions to Mars. Aldrin's vision extends beyond exploration to the establishment of permanent human settlements on the Red Planet, ensuring the survival and advancement of humanity.

In addition to his advocacy for Mars exploration, Aldrin has supported numerous educational initiatives aimed at inspiring young people to pursue careers in science, technology, engineering, and mathematics (STEM). He founded the ShareSpace Foundation, which promotes science literacy and education through hands-on experiences and educational programs. Aldrin's dedication to inspiring the next generation of explorers and scientists is a testament to his belief in the power of education and innovation.

Aldrin's contributions to space exploration have been recognized with numerous awards and honors. He received the Presidential Medal of Freedom, the Congressional Gold Medal, and the NASA Distinguished Service Medal, among many others. He was inducted into the International Space Hall of Fame, the National Aviation Hall of Fame, and the United States Astronaut Hall of Fame. These accolades reflect his significant impact on space exploration and his enduring legacy.

In popular culture, Aldrin's achievements and persona have been featured in various films, documentaries, and television shows. His character has been portrayed in movies such as "Apollo 13" (1995) and "First Man" (2018), highlighting his role in the Apollo 11 mission. Aldrin has also made cameo appearances in shows like "The Simpsons" and "The Big Bang Theory," further cementing his status as a cultural icon.

In recent years, Aldrin has continued to be an active and vocal advocate for space exploration. He has participated in numerous conferences, panel discussions, and media interviews, sharing his insights and vision for the future. His tireless efforts to promote space exploration and inspire future generations underscore his commitment to advancing human knowledge and achievement.

Buzz Aldrin's life and career are a testament to the power of human determination, innovation, and exploration. From his early days as a combat pilot in the Korean War to his historic moonwalk and his ongoing advocacy for Mars exploration, Aldrin has demonstrated an unwavering commitment to pushing the boundaries of what is possible. His contributions to space exploration have left an indelible mark on history, inspiring countless individuals to reach for the stars.

Aldrin's legacy extends beyond his achievements in space. His openness about his personal struggles and his advocacy for mental health awareness have made him a relatable and inspiring figure. His dedication to education and his efforts to inspire the next generation of scientists and explorers reflect his belief in the power of knowledge and innovation to shape a better future.

Chapter 36: Michael Collins

Michael Collins, a distinguished astronaut and a pivotal figure in the history of space exploration, is best known for his role as the command module pilot for Apollo 11, the first manned mission to land on the moon. Born on October 31, 1930, in Rome, Italy, where his father, a U.S. Army Major General, was stationed, Collins had a unique upbringing that exposed him to various cultures and experiences. His early years laid the foundation for a career marked by discipline, intelligence, and an unwavering commitment to exploration.

Collins' family returned to the United States when he was young, and he spent his childhood moving frequently due to his father's military career. He attended St. Albans School in Washington, D.C., where he excelled academically and participated in various extracurricular activities, including rowing. His disciplined upbringing and academic excellence earned him an appointment to the United States Military Academy at West Point, from which he graduated in 1952 with a degree in military science.

Following his graduation, Collins chose to join the United States Air Force, inspired by his fascination with flight and the desire to serve his country. He completed flight training at Columbus Air Force Base in Mississippi and went on to become an accomplished fighter pilot. Collins' early military career included assignments in the Philippines and at Nellis Air Force Base in Nevada, where he honed his skills in flying advanced aircraft. His proficiency as a pilot and his leadership qualities led to his selection for the U.S. Air Force Experimental Flight Test Pilot School at Edwards Air Force Base in California.

It was at Edwards that Collins' career took a significant turn towards space exploration. In 1963, he applied for and was accepted into NASA's third group of astronauts, known as the "Third Group" or "The Fourteen." This group was instrumental in advancing the United States' space program during the Gemini and Apollo missions. Collins' first spaceflight was as pilot of Gemini 10 in July 1966, alongside command pilot John Young. The mission's objectives included docking with an Agena target vehicle and conducting extravehicular activities (EVAs). Collins successfully performed two EVAs, demonstrating the feasibility of working in space and advancing the techniques needed for future lunar missions.

Collins' performance on Gemini 10 showcased his technical expertise and ability to handle the challenges of spaceflight. His calm demeanor, problem-solving skills, and physical endurance made him an ideal candidate for the demanding missions that lay ahead. In January 1969, he was selected as the command module pilot for Apollo 11, NASA's ambitious mission to land a man on the moon and return him safely to Earth. The crew also included mission commander Neil Armstrong and lunar module pilot Buzz Aldrin.

As the command module pilot, Collins was responsible for piloting the command module, Columbia, which would remain in lunar orbit while Armstrong and Aldrin descended to the moon's surface in the lunar module, Eagle. Collins played a crucial role in the mission, ensuring that the command module functioned flawlessly and was ready to rendezvous with the lunar module upon its return from the surface. His meticulous attention to detail and rigorous training were vital to the mission's success.

Apollo 11 launched from Kennedy Space Center on July 16, 1969, atop a Saturn V rocket. The journey to the moon took approximately three days, during which Collins maintained the command module and prepared for the critical phases of the mission. On July 20, 1969, while Armstrong and Aldrin made their historic descent to the moon's surface, Collins remained alone in the command module, orbiting the moon. For nearly 21 hours, he circled the moon, conducting experiments, taking photographs, and monitoring the systems of Columbia.

Collins' time alone in lunar orbit is often described as one of the most solitary experiences in human history. He later reflected on the profound sense of isolation and responsibility he felt, knowing that he was the sole lifeline for his fellow astronauts on the lunar surface. Despite the solitude, Collins found the experience deeply meaningful and spiritually uplifting. He marveled at the beauty of the moon and the Earth from space, gaining a unique perspective on humanity's place in the universe.

The successful completion of the lunar landing and the safe return of Armstrong and Aldrin to the command module marked a triumphant moment in human history. Collins' precise piloting and unwavering focus were instrumental in ensuring the mission's success. On July 24, 1969, Apollo 11 splashed down in the Pacific Ocean, and the astronauts were celebrated worldwide as heroes.

Following Apollo 11, Collins chose not to fly in space again. Instead, he took on significant roles within NASA and beyond. He served as Assistant Secretary of State for Public Affairs, where he played a key role in promoting the achievements of the U.S. space program and fostering international cooperation in space exploration. In 1971, he became the director of the National Air and Space Museum in Washington, D.C., overseeing its development into one of the world's premier institutions dedicated to aviation and space history.

During his tenure at the museum, Collins was instrumental in planning and executing the museum's grand opening in 1976, which coincided with the United States Bicentennial celebrations. His leadership and vision helped transform the museum into a world-class institution that continues to inspire and educate millions of visitors each year. Collins' contributions to preserving and promoting the history of flight and space exploration have had a lasting impact on public understanding and appreciation of these fields.

In addition to his work at the museum, Collins authored several books, including his autobiography "Carrying the Fire" (1974), which is widely regarded as one of the best astronaut memoirs ever written. His other works include "Flying to the Moon: An Astronaut's Story" (1976) and "Liftoff: The Story of America's Adventure in Space" (1988). Through his writing, Collins shared his experiences and insights, providing a firsthand account of the challenges and triumphs of space exploration.

Collins retired from the Air Force with the rank of Major General in 1982. After leaving the National Air and Space Museum, he continued to be involved in various aerospace and business ventures. He served on several corporate boards and was a sought-after speaker, sharing his experiences and perspectives on space exploration, technology, and leadership.

Throughout his life, Collins remained a passionate advocate for space exploration. He emphasized the importance of continuing human missions beyond low Earth orbit and exploring Mars and other destinations in the solar system. Collins believed that the spirit of exploration and the pursuit of knowledge were fundamental to humanity's progress and survival. His advocacy efforts included participating in conferences, interviews, and public events, where he inspired new generations of scientists, engineers, and explorers.

Collins' legacy extends beyond his achievements in spaceflight. He was known for his humility, intellectual curiosity, and dedication to public service. His contributions to space exploration, education, and the preservation of aviation history have left an indelible mark on the field and continue to inspire and inform future generations.

Collins passed away on April 28, 2021, at the age of 90. His death was met with an outpouring of tributes from around the world, reflecting his profound impact on space exploration and his enduring legacy. NASA Administrator Bill Nelson described Collins as "one of America's greatest explorers," highlighting his contributions to the Apollo 11 mission and his lifelong commitment to advancing human spaceflight.

Chapter 37: Jim Lovell

James Arthur "Jim" Lovell Jr., born on March 25, 1928, in Cleveland, Ohio, is one of NASA's most distinguished astronauts and a key figure in the history of space exploration. Best known for his role as the commander of the ill-fated Apollo 13 mission, Lovell's career is marked by his exceptional leadership, his significant contributions to manned spaceflight, and his resilience in the face of adversity. His story is one of perseverance, technical expertise, and the human spirit's enduring quest for discovery.

Lovell's early life was marked by a fascination with aviation and exploration. His interest in rocketry began in his youth, inspired by Robert Goddard's pioneering work in rocketry and the science fiction tales of Jules Verne. He attended Juneau High School in Milwaukee, Wisconsin, where his passion for flight was further nurtured. After high school, he attended the University of Wisconsin–Madison for two years before transferring to the United States Naval Academy, from which he graduated in 1952 with a degree in naval science.

Upon graduating from the Naval Academy, Lovell entered flight training and became a naval aviator. His early career was distinguished by his skill and dedication, flying a variety of aircraft and serving aboard carriers. He also attended the United States Naval Test Pilot School at Patuxent River, Maryland, which honed his skills in aviation and prepared him for his future role as an astronaut.

In 1962, Lovell was selected as part of NASA's second group of astronauts, known as the "New Nine." This group included notable figures such as Neil Armstrong and Frank Borman, with whom Lovell would later fly in space. His selection marked the beginning of a remarkable career in space exploration.

Lovell's first spaceflight was aboard Gemini 7 in December 1965, where he served as the pilot alongside commander Frank Borman. The mission was a significant milestone in NASA's Gemini program, as it involved a 14-day endurance flight, the longest spaceflight at that time. Gemini 7 also accomplished the first successful space rendezvous with Gemini 6A, commanded by Wally Schirra. The mission provided critical data on the effects of long-duration spaceflight on the human body, which was essential for planning future Apollo missions to the Moon.

Lovell's performance on Gemini 7 earned him a reputation as a capable and reliable astronaut. He returned to space in November 1966 as the command pilot of Gemini 12, the final mission of the Gemini program. His crewmate on this mission was Edwin "Buzz" Aldrin, who would later become one of the first two humans to walk on the Moon. During Gemini 12, Lovell and Aldrin conducted a series of complex extravehicular activities (EVAs), or spacewalks, which demonstrated improved techniques for working in space. The mission also included docking maneuvers with an Agena target vehicle, further advancing NASA's capabilities in orbital operations.

Following the success of the Gemini program, Lovell was assigned to the Apollo program, which aimed to land humans on the Moon and return them safely to Earth. His first Apollo flight was Apollo 8 in December 1968, a mission that made history as the first manned spacecraft to leave Earth orbit, travel to the Moon, and return. Alongside Frank Borman and Bill Anders, Lovell orbited the Moon and provided critical reconnaissance of potential landing sites for future missions. Apollo 8's iconic "Earthrise" photograph, taken by Anders, became a symbol of the mission and a powerful image of the Earth from space. The mission's success was a significant step toward achieving the goal of a lunar landing and demonstrated the capabilities of the Saturn V rocket and the Apollo spacecraft.

Lovell's next and most famous mission was Apollo 13, launched on April 11, 1970. As the mission commander, Lovell was joined by command module pilot Jack Swigert and lunar module pilot Fred Haise. Apollo 13 was intended to be the third manned lunar landing, but the mission was dramatically altered when an oxygen tank in the service module exploded two days into the flight. The explosion severely damaged the spacecraft, leading to the loss of most of the crew's oxygen and electrical power. The iconic phrase "Houston, we've had a problem" (often misquoted as "Houston, we have a problem") communicated the gravity of the situation to mission control.

Faced with a life-threatening crisis, Lovell and his crew, with the support of NASA's ground team, worked tirelessly to devise and execute a plan to bring the spacecraft safely back to Earth. They used the lunar module, Aquarius, as a "lifeboat" to provide life support and propulsion. The crew endured extreme cold, limited power, and rising levels of carbon dioxide. Lovell's leadership and calm demeanor under pressure were crucial in maintaining the crew's focus and morale. Innovative problem-solving and teamwork allowed the astronauts to make a successful reentry and splashdown in the Pacific Ocean on April 17, 1970. The mission, although a failure in its primary objective, was a triumph of human ingenuity and determination, and it further cemented Lovell's status as a hero.

Apollo 13's successful return was celebrated worldwide, and Lovell received numerous accolades for his role in the mission. He was awarded the Presidential Medal of Freedom, the United States' highest civilian honor, in recognition of his extraordinary efforts. The mission also inspired the 1995 film "Apollo 13," in which actor Tom Hanks portrayed Lovell, bringing the story of the mission to a new generation.

Following Apollo 13, Lovell continued to contribute to NASA and the space program. He was slated to command Apollo 14 but was reassigned after the mission's crew was reshuffled due to medical concerns. Instead, he transitioned to a managerial role, serving as the Deputy Director of the Johnson Space Center. Lovell retired from the Navy and NASA in 1973 with the rank of Captain, ending a distinguished career in both organizations.

After leaving NASA, Lovell entered the private sector, holding executive positions at various companies, including Bay-Houston Towing Company and Fisk Telephone Systems. He also co-authored a memoir with journalist Jeffrey Kluger, titled "Lost Moon: The Perilous Voyage of Apollo 13," which provided an in-depth account of the mission. The book served as the basis for the film "Apollo 13," further solidifying Lovell's legacy in popular culture.

Throughout his life, Lovell has remained an advocate for space exploration and education. He has participated in numerous speaking engagements, sharing his experiences and insights with audiences around the world. Lovell's dedication to inspiring future generations of scientists, engineers, and explorers reflects his enduring passion for discovery and innovation.

In recognition of his contributions to space exploration, Lovell has received numerous honors and awards. In addition to the Presidential Medal of Freedom, he has been inducted into the U.S. Astronaut Hall of Fame, the National Aviation Hall of Fame, and the International Space Hall of Fame. He has also received honorary degrees from several universities and has been honored by various scientific and professional organizations.

Lovell's legacy extends beyond his technical achievements and heroic deeds. His story embodies the spirit of exploration and the pursuit of knowledge, demonstrating the importance of perseverance, teamwork, and leadership in overcoming challenges. His experiences during Apollo 13, in particular, highlight the resilience and ingenuity required to navigate crises and achieve success in the face of adversity.

Jim Lovell's impact on space exploration and his role as a pioneer of human spaceflight continue to inspire new generations of space enthusiasts and professionals. His contributions have helped to pave the way for future missions and advancements in space technology, ensuring that humanity's journey into the cosmos will continue for years to come.

As we look to the future of space exploration, with plans for returning to the Moon and eventually reaching Mars, Lovell's legacy serves as a guiding light. His dedication to pushing the boundaries of human capability and his commitment to exploring the unknown remain central to the ethos of space exploration. Jim Lovell's life and career are a testament to the extraordinary achievements that can be realized through vision, courage, and a relentless pursuit of knowledge.

Chapter 38: Valentina Tereshkova

Valentina Vladimirovna Tereshkova, born on March 6, 1937, in the village of Maslennikovo in the Yaroslavl Oblast of Russia, is renowned as the first woman to have flown in space. Her historic mission aboard Vostok 6 on June 16, 1963, marked a significant milestone in human space exploration and the broader struggle for gender equality in a field traditionally dominated by men. Tereshkova's life story is one of remarkable perseverance, ambition, and groundbreaking achievements that transcended her role as a cosmonaut.

Tereshkova was born into a family of humble means; her father was a tractor driver and her mother worked in a textile plant. Her father died during World War II, leaving her mother to raise three children alone. Despite these challenges, Tereshkova exhibited a strong work ethic and a thirst for knowledge from an early age. She attended school intermittently due to family obligations but managed to complete her education through correspondence courses while working to support her family.

Her passion for aviation and skydiving began during her youth. Tereshkova joined a local aero club and became an avid parachutist, making her first jump at the age of 22. Her skill and enthusiasm for skydiving led to her selection as a cosmonaut. In 1961, the Soviet Union launched Yuri Gagarin into space, making him the first human to orbit Earth. Inspired by this achievement, Soviet Premier Nikita Khrushchev sought to continue the Soviet Union's dominance in space by sending a woman into space.

In 1962, the Soviet space program began recruiting female cosmonauts, and Tereshkova was among the five women selected for training. Despite having no formal pilot training, her extensive experience in skydiving made her a strong candidate. The rigorous training regimen included weightless flights, isolation tests, centrifuge tests to gauge her reactions to high g-forces, and numerous academic courses on rocket theory, spacecraft engineering, and astronautical medicine.

After rigorous training and selection processes, Tereshkova was chosen to pilot Vostok 6. On June 16, 1963, she launched from the Baikonur Cosmodrome in Kazakhstan, becoming not only the first woman in space but also the first civilian to fly into space, as she was not a member of the military. Her call sign for the mission was "Chaika," meaning "Seagull," a symbol of her graceful and pioneering flight.

The mission lasted nearly three days, during which Tereshkova orbited Earth 48 times. Throughout her mission, she conducted various experiments, took photographs of the Earth, and maintained a detailed log of her experiences. Despite experiencing physical discomfort and nausea, she completed her mission objectives with determination and resilience. Her successful return to Earth on June 19, 1963, was a momentous occasion, celebrated by the Soviet Union and heralded around the world as a triumph for women in space.

Tereshkova's achievement had a profound impact on the Soviet space program and on the perception of women's capabilities in science and technology. She was awarded numerous honors, including the title of Hero of the Soviet Union, the highest distinction in the USSR. Her flight became a symbol of Soviet progressiveness and the country's commitment to gender equality, bolstering its image during the Cold War.

Following her historic flight, Tereshkova continued to play an active role in the Soviet space program. She served as a spokesperson and advocate for space exploration, traveling extensively to promote Soviet achievements and inspire future generations. She also pursued higher education, earning a degree in engineering from the Zhukovsky Air Force Engineering Academy in 1969, and later a doctorate in aeronautical engineering.

Tereshkova's career extended beyond her contributions to spaceflight. She became involved in politics, serving as a member of the Supreme Soviet, the highest legislative body in the Soviet Union, from 1966 to 1991. She was also a member of the Central Committee of the Communist Party and held various positions within the party, including head of the Soviet Women's Committee. Her political career was marked by her advocacy for women's rights, social welfare, and scientific advancement.

In addition to her political work, Tereshkova remained a prominent figure in international space organizations. She was an active member of the International Association of Space Explorers and participated in various scientific conferences and symposiums. Her contributions to space exploration and her efforts to promote international cooperation in space science earned her numerous accolades and recognition from countries around the world.

Tereshkova's legacy extends beyond her pioneering spaceflight. Her story has inspired countless women to pursue careers in science, technology, engineering, and mathematics (STEM). She broke barriers and challenged stereotypes, proving that women could excel in the demanding field of space exploration. Her accomplishments have been celebrated in books, documentaries, and films, ensuring that her contributions to space exploration are remembered and honored.

In her later years, Tereshkova continued to be an influential figure in Russia. She remained active in politics, serving as a deputy in the State Duma, the lower house of the Russian parliament, where she focused on issues related to space exploration, science, and technology. Her dedication to public service and her ongoing advocacy for space exploration highlight her enduring commitment to advancing human knowledge and exploration.

Valentina Tereshkova's life is a testament to the power of determination, resilience, and the pursuit of dreams. From her humble beginnings in a small Russian village to her historic journey into space and her influential political career, she has left an indelible mark on the history of space exploration and women's rights. Her pioneering spirit and her contributions to science and technology continue to inspire new generations of explorers, scientists, and engineers.

As we look to the future of space exploration, Tereshkova's legacy serves as a reminder of the importance of inclusivity and diversity in achieving great advancements. Her journey from a textile worker to a space pioneer demonstrates that with determination, passion, and support, anyone can reach for the stars. Her story is a celebration of human potential and the boundless possibilities that lie ahead in our quest to explore the universe.

Chapter 39: Shannon Lucid

Shannon Matilda Wells Lucid, born on January 14, 1943, in Shanghai, China, is an American biochemist and retired NASA astronaut renowned for her groundbreaking achievements in space exploration. Her career is distinguished by numerous records and honors, reflecting her significant contributions to space science and her role as a trailblazer for women in the field of astronautics. Lucid's journey from a young girl fascinated by science to one of NASA's most accomplished astronauts is a story of perseverance, dedication, and extraordinary achievement.

Lucid's early life was shaped by unique experiences. Born to Baptist missionary parents, she spent the early years of her life in China. Her family was interned by the Japanese during World War II, but they returned to the United States in 1945. The Lucids settled in Bethany, Oklahoma, where Shannon developed a keen interest in science. Her passion for exploring the unknown was evident from a young age, influenced by her parents' missionary work and the stories they shared about their experiences.

Lucid pursued her education with determination. She graduated from Bethany High School in 1960 and went on to attend the University of Oklahoma, where she earned a Bachelor of Science degree in chemistry in 1963. She continued her studies at the same institution, obtaining a Master of Science degree in biochemistry in 1970 and a Ph.D. in biochemistry in 1973. Her doctoral research focused on the structure and function of proteins, laying a solid foundation for her future work in space science.

Lucid's path to becoming an astronaut began with her work as a scientist. After completing her Ph.D., she worked as a research chemist at the Oklahoma Medical Research Foundation and later as a research associate at the University of Oklahoma Health Sciences Center. Her expertise in biochemistry and her passion for space exploration made her an ideal candidate for NASA's astronaut program, which was increasingly seeking scientists to conduct research in space.

In 1978, Lucid was selected as one of the first six women to join NASA's astronaut corps. This selection was part of NASA's effort to diversify its astronaut ranks and include more scientists and engineers, reflecting a broader range of expertise and perspectives. Lucid and her fellow female astronauts, including Sally Ride and Judith Resnik, were pioneers who broke new ground for women in the field of space exploration.

Lucid's first spaceflight was on STS-51-G, a mission of the Space Shuttle Discovery, launched on June 17, 1985. During this mission, she served as a mission specialist, conducting experiments and deploying satellites. The mission lasted nearly eight days and demonstrated the versatility of the Space Shuttle for conducting a wide range of scientific and technical tasks in space. Lucid's performance on this mission established her as a capable and reliable astronaut, paving the way for future assignments.

Her second spaceflight was on STS-34, aboard the Space Shuttle Atlantis, launched on October 18, 1989. This mission was notable for deploying the Galileo spacecraft, which was designed to study Jupiter and its moons. Lucid's role as a mission specialist involved overseeing the deployment of Galileo and conducting scientific experiments. The mission highlighted her expertise in handling complex scientific instruments and contributed to our understanding of the Jovian system.

Lucid's third spaceflight was on STS-43, aboard the Space Shuttle Atlantis, launched on August 2, 1991. This mission focused on deploying the Tracking and Data Relay Satellite (TDRS-E) and conducting various scientific experiments. Lucid's work on this mission further demonstrated her proficiency in space operations and her ability to manage intricate scientific tasks in the challenging environment of space.

Her fourth spaceflight was on STS-58, aboard the Space Shuttle Columbia, launched on October 18, 1993. This mission was dedicated to life sciences research, and Lucid served as a payload commander. The mission's primary objective was to study the effects of microgravity on the human body, using a variety of biological experiments. Lucid's background in biochemistry made her an ideal candidate for this mission, and her contributions significantly advanced our understanding of how spaceflight impacts human physiology.

Lucid's fifth and most notable spaceflight was on the Russian space station Mir. In March 1996, she launched aboard the Space Shuttle Atlantis on STS-76 and spent 188 days on Mir, setting a record for the longest duration spaceflight by a woman and an American astronaut at the time. Her mission involved conducting a wide range of scientific experiments, including studies on the effects of long-duration spaceflight on the human body, materials science research, and various biological experiments. Lucid's stay on Mir was part of the Shuttle-Mir program, a collaborative effort between NASA and the Russian space agency to foster international cooperation in space exploration.

During her time on Mir, Lucid faced numerous challenges, including the psychological and physical demands of long-duration spaceflight. She adapted to the confined living conditions, developed strong working relationships with her Russian colleagues, and conducted valuable scientific research. Her resilience and dedication were widely recognized, and she became a symbol of international cooperation in space. Lucid's mission on Mir also provided critical insights and experience that contributed to the development of the International Space Station (ISS).

Lucid's achievements in space were recognized with numerous awards and honors. She received the Congressional Space Medal of Honor in 1996, one of the highest honors bestowed upon astronauts for their contributions to space exploration. In addition, she was inducted into the U.S. Astronaut Hall of Fame and received several

NASA awards, including the NASA Distinguished Service Medal and the NASA Exceptional Service Medal.

Following her historic mission on Mir, Lucid continued to play an active role in NASA. She served as the Chief Scientist of NASA from 2002 to 2003, where she provided leadership and guidance on scientific research and policy. Her contributions to NASA's scientific programs and her advocacy for the importance of scientific research in space exploration were highly influential.

Lucid retired from NASA in 2012, leaving behind a legacy of remarkable achievements and groundbreaking contributions to space exploration. Her career serves as an inspiration to countless individuals, particularly women and girls interested in pursuing careers in science, technology, engineering, and mathematics (STEM). Lucid's story is a testament to the power of perseverance, dedication, and the pursuit of knowledge.

Throughout her career, Lucid demonstrated an unwavering commitment to advancing our understanding of space and the human body's response to spaceflight. Her contributions to space science, particularly in the field of biochemistry and life sciences, have had a lasting impact on the scientific community and on the development of future space missions. Her work on the Shuttle-Mir program and her leadership in scientific research have paved the way for continued international collaboration in space exploration.

In addition to her scientific and technical achievements, Lucid's legacy includes her role as a trailblazer for women in space. She broke barriers and challenged stereotypes, proving that women could excel in the demanding field of astronautics. Her success has inspired generations of women to pursue careers in science and engineering, and her story continues to motivate aspiring astronauts and scientists around the world.

Lucid's life and career are a testament to the power of curiosity, determination, and the human spirit's enduring quest for discovery.

Her contributions to space exploration and her dedication to advancing scientific knowledge have left an indelible mark on the history of astronautics. As we look to the future of space exploration, Shannon Lucid's legacy serves as a guiding light, reminding us of the incredible potential of human achievement and the importance of fostering diversity and inclusivity in our pursuit of the stars.

Chapter 40: Yuri Gagarin

Yuri Alekseyevich Gagarin, born on March 9, 1934, in the village of Klushino near Gzhatsk (now Gagarin) in the Soviet Union, is one of the most celebrated figures in the history of space exploration. He is best known for being the first human to journey into outer space, a monumental achievement that marked the beginning of human spaceflight. Gagarin's life story is a remarkable tale of courage, determination, and extraordinary achievement, set against the backdrop of the intense geopolitical rivalry of the Cold War era.

Gagarin was born into a humble family; his father was a carpenter and his mother was a milkmaid. Despite the modest circumstances of his upbringing, he displayed an early interest in aviation and science. The hardships of World War II, including the German occupation of his village, did not deter young Gagarin's ambitions. After the war, the family moved to Gzhatsk, and Gagarin continued his education, excelling in mathematics and physics.

In 1951, Gagarin enrolled at the Saratov Industrial Technical School, where he studied to become a foundryman. During his time in Saratov, he joined a local flying club and made his first solo flight in a Yak-18 trainer aircraft. This experience solidified his passion for aviation, and he decided to pursue a career as a pilot. In 1955, he was accepted into the First Chkalovsky Higher Air Force Pilots School in Orenburg, where he trained to become a fighter pilot. Gagarin graduated in 1957 and was commissioned as a lieutenant in the Soviet Air Force.

Gagarin's skill as a pilot, combined with his exceptional physical fitness and strong work ethic, caught the attention of Soviet space program officials. In 1960, he was selected as one of the original 20 cosmonauts for the Soviet space program, a prestigious and highly competitive position. The selection process was rigorous, involving extensive physical and psychological testing. Gagarin's exemplary performance in these tests, along with his charismatic personality, made him a standout candidate.

On April 12, 1961, Yuri Gagarin made history as the first human to journey into outer space. He was launched aboard the Vostok 1 spacecraft from the Baikonur Cosmodrome in Kazakhstan. The mission, which lasted 108 minutes, took Gagarin into orbit around the Earth, reaching a maximum altitude of 327 kilometers (about 203 miles). As the spacecraft orbited the Earth, Gagarin experienced weightlessness and witnessed the planet from a vantage point that no human had ever seen before. His calm demeanor and professional conduct during the flight were remarkable, considering the unprecedented nature of the mission and the inherent risks involved.

The flight of Vostok 1 was a major triumph for the Soviet space program and a significant propaganda victory during the Cold War. Gagarin's successful mission demonstrated the Soviet Union's technological prowess and underscored its leadership in space exploration. Upon his return, Gagarin was hailed as a national hero and became an international symbol of Soviet achievement. He received numerous awards and honors, including the title of Hero of the Soviet Union, the highest distinction in the country.

Gagarin's achievement had a profound impact on the space race, intensifying the competition between the United States and the Soviet Union. His flight prompted the United States to accelerate its own space program, leading to the eventual moon landing in 1969. Gagarin's mission also inspired millions of people around the world and fueled public interest in space exploration.

Following his historic flight, Gagarin embarked on a global tour, visiting numerous countries as a goodwill ambassador for the Soviet Union. His charisma, humility, and enthusiasm for space exploration made him a beloved figure internationally. Gagarin's public appearances and speeches emphasized the peaceful nature of space exploration and the potential for international cooperation in scientific endeavors.

Despite his newfound celebrity status, Gagarin continued to pursue his passion for aviation and space. He resumed his duties with the Soviet Air Force and remained involved in the space program. In 1962, he was appointed deputy training director of the Cosmonaut Training Center, where he played a key role in preparing future cosmonauts for space missions. Gagarin's experience and insights were invaluable in shaping the training protocols and safety measures for subsequent spaceflights.

Gagarin also continued his education, enrolling at the Zhukovsky Air Force Engineering Academy. He graduated in 1968 with a degree in aeronautical engineering, further enhancing his technical expertise. His commitment to continuous learning and professional development reflected his dedication to advancing the field of space exploration.

Tragically, Yuri Gagarin's life was cut short on March 27, 1968, when he died in a plane crash during a routine training flight near the town of Kirzhach. The cause of the crash has been the subject of much speculation and investigation, but it is generally believed to have been an accident resulting from a combination of factors, including adverse weather conditions and pilot error. Gagarin's untimely death was a profound loss for the Soviet Union and the global space community.

Gagarin's legacy, however, endures. His pioneering flight into space remains a landmark achievement in human history, symbolizing the spirit of exploration and the quest for knowledge. Gagarin's name is commemorated in numerous ways, including the renaming of his hometown to Gagarin, the establishment of Gagarin's Cosmonaut Training Center, and various monuments and memorials around the world. The annual Yuri's Night, celebrated on April 12, honors his historic flight and promotes space exploration.

In addition to his contributions to space exploration, Gagarin's life story serves as an inspiration for aspiring astronauts, scientists, and engineers. His journey from a rural village to the cosmos exemplifies the potential for individuals to achieve extraordinary things through determination, hard work, and the pursuit of their passions. Gagarin's achievement also highlights the importance of international cooperation in space exploration, as his mission paved the way for collaborative efforts such as the International Space Station.

Gagarin's impact extends beyond his historical flight. He played a crucial role in shaping the Soviet space program and influencing the development of space exploration policies. His insights and experiences informed the design and implementation of subsequent missions, contributing to the advancement of human spaceflight capabilities. Gagarin's emphasis on the peaceful exploration of space and the potential for scientific discovery continues to resonate in contemporary discussions about the future of space exploration.

In recognition of his contributions, Gagarin has been honored with numerous awards and accolades posthumously. His legacy is celebrated in various fields, including education, where institutions and scholarships bear his name, encouraging future generations to pursue careers in science and technology. Gagarin's influence also extends to popular culture, with his story being depicted in films, books, and documentaries, ensuring that his achievements are remembered and appreciated by a wide audience.

Yuri Gagarin's life and career embody the spirit of human exploration and the relentless pursuit of knowledge. His historic flight into space was a monumental achievement that not only marked a significant milestone in the history of space exploration but also inspired generations of scientists, engineers, and dreamers to reach for the stars. Gagarin's legacy is a testament to the power of human ingenuity and the enduring quest to explore the unknown, reminding us of the incredible potential that lies within each individual to achieve greatness.

Chapter 41: Eugene Cernan

Eugene Andrew Cernan, born on March 14, 1934, in Chicago, Illinois, was an American astronaut, naval aviator, electrical engineer, aeronautical engineer, and fighter pilot. He is best known for his role as the commander of Apollo 17, the final mission of NASA's Apollo program, during which he became the last human to walk on the lunar surface. Cernan's life and career are a testament to the spirit of exploration, dedication, and pioneering achievement in space exploration.

Cernan grew up in the Chicago suburbs and displayed an early interest in aviation and engineering. He attended Proviso East High School in Maywood, Illinois, where he excelled academically and athletically. After graduating high school in 1952, Cernan pursued higher education at Purdue University, a prestigious institution known for its engineering programs. He graduated in 1956 with a Bachelor of Science degree in Electrical Engineering. His passion for aviation led him to join the Naval Reserve Officers Training Corps (NROTC) at Purdue, which set the stage for his future career as a naval aviator and astronaut.

Following his graduation, Cernan was commissioned as an ensign in the United States Navy. He underwent flight training and earned his Naval Aviator Wings in 1957. Cernan then served as a fighter pilot, flying FJ-4 Fury and A-4 Skyhawk aircraft. His skill and dedication as a pilot earned him a reputation as a top-notch aviator. In addition to his flying duties, Cernan continued his education, earning a Master of Science degree in Aeronautical Engineering from the U.S. Naval Postgraduate School in Monterey, California, in 1963.

In 1963, Cernan was selected by NASA as part of the third group of astronauts, often referred to as the "New Nine." This group included notable astronauts such as Buzz Aldrin, Michael Collins, and Alan Bean. Cernan's selection came at a time when NASA was intensifying its efforts to achieve President John F. Kennedy's goal of landing a man on the Moon and returning him safely to Earth by the end of the decade.

Cernan's first spaceflight was as the pilot of Gemini 9A, which launched on June 3, 1966. The mission was initially intended to include a space rendezvous and docking with an Agena target vehicle. However, due to a malfunction, the Agena vehicle was replaced by an alternative target, the Augmented Target Docking Adapter (ATDA). During the mission, Cernan became the second American to perform a spacewalk (extravehicular activity or EVA), spending over two hours outside the spacecraft. The spacewalk proved to be extremely challenging due to the lack of handholds and foot restraints, causing Cernan to expend a significant amount of energy. Despite these difficulties, the mission provided valuable data and experience for future spacewalks.

Cernan's second spaceflight was as the lunar module pilot of Apollo 10, which launched on May 18, 1969. Apollo 10 was a critical "dress rehearsal" for the first Moon landing, designed to test all aspects of the lunar landing mission except for the actual landing itself. Cernan, along with commander Thomas Stafford and command module pilot John Young, descended to within 8.4 nautical miles of the lunar surface in the lunar module, named "Snoopy." The mission successfully demonstrated the performance of the lunar module in the Moon's gravitational field and provided valuable experience in lunar orbit operations. Apollo 10's success paved the way for the historic Apollo 11 mission, which achieved the first manned Moon landing just two months later.

Cernan's most significant achievement came with his third and final spaceflight as the commander of Apollo 17, the last mission of NASA's Apollo program. Apollo 17 launched on December 7, 1972, with Cernan, lunar module pilot Harrison Schmitt, and command module pilot Ronald Evans on board. The mission aimed to conduct extensive scientific exploration of the lunar surface and included the first professional geologist, Schmitt, to walk on the Moon. Apollo 17 was also notable for being the first night launch of the Saturn V rocket.

Upon reaching the Moon, Cernan and Schmitt conducted three moonwalks, spending a total of 22 hours and 4 minutes outside the lunar module. They explored the Taurus-Littrow valley, a region selected for its diverse geological features, which included highlands, valleys, and volcanic formations. The astronauts collected over 110 kilograms (243 pounds) of lunar samples, conducted scientific experiments, and deployed scientific instruments. One of the most famous images from the mission is the "Blue Marble" photograph of Earth, taken by the crew during their journey to the Moon.

Cernan's final words before leaving the lunar surface on December 14, 1972, were poignant and reflective: "As I take man's last step from the surface, back home for some time to come, but we believe not too long into the future—I'd like to just say what I believe history will record: that America's challenge of today has forged man's destiny of tomorrow. And, as we leave the Moon at Taurus-Littrow, we leave as we came, and, God willing, as we shall return, with peace and hope for all mankind. Godspeed the crew of Apollo 17."

These words encapsulated the spirit of the Apollo program and humanity's enduring quest to explore and understand the universe. Cernan's leadership, bravery, and contributions to space exploration left an indelible mark on the history of spaceflight.

Following his retirement from NASA and the Navy in 1976, Cernan pursued a career in business and public speaking. He served as an executive for Coral Petroleum, Inc., and later founded his own aerospace consulting firm, The Cernan Corporation. He was also involved in various ventures related to space and technology, including serving as a television commentator for space shuttle launches and as a motivational speaker.

Cernan remained an advocate for space exploration throughout his life, often speaking about the importance of continuing human spaceflight missions beyond low Earth orbit. He was a vocal proponent of returning to the Moon and eventually sending humans to Mars. His experiences and insights as an astronaut provided valuable perspectives on the challenges and opportunities of space exploration.

In addition to his professional accomplishments, Cernan was recognized with numerous awards and honors. He received the NASA Distinguished Service Medal, the Navy Distinguished Flying Cross, and the Navy Distinguished Service Medal, among others. Cernan was also inducted into the U.S. Astronaut Hall of Fame and the International Space Hall of Fame.

Cernan's legacy extends beyond his space missions. He authored an autobiography, "The Last Man on the Moon," co-written with journalist Don Davis, which was published in 1999. The book provides a candid and personal account of his life, his experiences as an astronaut, and his reflections on the Apollo program. The book was later adapted into a documentary film, released in 2016, which brought Cernan's story to a wider audience.

Throughout his life, Cernan remained a source of inspiration for future generations of explorers, scientists, and engineers. His dedication to space exploration, his willingness to take risks, and his commitment to advancing human knowledge exemplify the qualities of a true pioneer. Cernan's contributions to the Apollo program and his role in one of humanity's greatest achievements continue to be celebrated and remembered.

Eugene Cernan passed away on January 16, 2017, at the age of 82. His death was widely mourned by the space community and the public, who recognized his significant contributions to space exploration and his enduring legacy. Cernan's life and career serve as a powerful reminder of the importance of exploration, the pursuit of knowledge, and the human spirit's capacity to achieve remarkable feats.

As we look to the future of space exploration, Cernan's words and actions continue to inspire us. His belief in the potential of space exploration to forge humanity's destiny and bring peace and hope to all mankind resonates with the ongoing efforts to return to the Moon, explore Mars, and venture further into the cosmos. Cernan's legacy is a testament to the power of human ingenuity, determination, and the unwavering desire to explore the unknown.

Chapter 42: Carolyn Porco

Carolyn C. Porco, born on March 6, 1953, in New York City, is an esteemed American planetary scientist, known for her pioneering work in the study of the outer planets and their moons. She is particularly recognized for her significant contributions to the exploration of the Saturn system through her leadership in the imaging science team of the Cassini mission. Porco's career is a testament to her exceptional skills as a scientist, her dedication to space exploration, and her ability to communicate the wonders of the cosmos to the public.

Porco's fascination with space began at a young age. Growing up in the 1960s, she was inspired by the space race and the Apollo missions, which captured the imagination of people around the world. Her interest in science and the universe led her to pursue a career in astronomy and planetary science. Porco attended the State University of New York at Stony Brook, where she earned a Bachelor of Science degree in Physics and Astronomy in 1974. She then went on to earn her Ph.D. in Planetary Science from the California Institute of Technology (Caltech) in 1983, where she conducted her doctoral research under the supervision of famed planetary scientist James Elliot.

Porco's early research focused on the rings of Saturn and the dynamics of planetary ring systems. Her work on the dynamics of Saturn's rings, including the discovery and analysis of wave patterns in the rings, was groundbreaking. She utilized data from the Pioneer and Voyager missions to study the intricate structures and interactions within Saturn's rings, advancing our understanding of these complex systems. Her research contributed to the field of planetary science by providing insights into the physical processes that govern the behavior of ring systems around giant planets.

In the early 1980s, Porco joined the Voyager imaging team, where she played a crucial role in analyzing the data returned by the Voyager 1 and Voyager 2 spacecraft during their historic flybys of the outer planets. Her work on the Voyager missions included studying the ring systems of Jupiter, Saturn, Uranus, and Neptune, as well as the moons of these planets. The Voyager missions provided an unprecedented view of the outer solar system, and Porco's contributions were instrumental in interpreting the data and making significant discoveries about these distant worlds.

One of Porco's notable achievements during the Voyager mission was her work on the discovery of the Neptune ring arcs. The Voyager 2 spacecraft, during its flyby of Neptune in 1989, revealed that the planet's rings were not continuous but instead consisted of discrete arcs. This discovery challenged existing theories about ring dynamics and prompted further investigation into the mechanisms that could create and maintain such structures. Porco's analysis of the Voyager data helped to unravel the mystery of Neptune's ring arcs and advanced our understanding of planetary ring dynamics.

In the 1990s, Porco became involved with the Cassini-Huygens mission, a collaborative project between NASA, the European Space Agency (ESA), and the Italian Space Agency (ASI) aimed at exploring Saturn and its complex system of rings and moons. Launched in 1997, the Cassini spacecraft arrived at Saturn in 2004 and began a detailed examination of the planet and its surroundings. Porco served as the leader of the Cassini Imaging Science Team, overseeing the design, operation, and analysis of the mission's imaging experiments.

Under Porco's leadership, the Cassini Imaging Team made numerous groundbreaking discoveries that transformed our understanding of the Saturn system. One of the most significant findings was the discovery of active geysers on the moon Enceladus. Cassini's images revealed plumes of water vapor and icy particles erupting from fractures near the south pole of Enceladus, indicating the presence of subsurface liquid water. This discovery had profound implications for the possibility of life beyond Earth, as it suggested that Enceladus could harbor a habitable environment beneath its icy crust.

Porco and her team also provided detailed images and analysis of Saturn's rings, revealing intricate structures and dynamic processes that had never been seen before. They observed the interactions between the rings and Saturn's moons, uncovering phenomena such as the propeller-like features caused by moonlets embedded within the rings. The high-resolution images captured by Cassini allowed scientists to study the fine details of ring particles and their behavior, deepening our understanding of the complex and dynamic nature of ring systems.

Another major achievement of the Cassini Imaging Team was the study of Saturn's largest moon, Titan. Cassini's images of Titan revealed a world with a dense atmosphere, lakes of liquid methane and ethane, and vast dune fields. The mission's findings on Titan provided valuable insights into the processes shaping its surface and atmosphere, and suggested that Titan, like Enceladus, could potentially support life. The detailed images and data collected by Cassini helped to build a comprehensive picture of Titan as a complex and diverse world with many Earth-like features.

Porco's work on the Cassini mission also included the study of Saturn's atmospheric phenomena, such as the planet's massive storms and the mysterious hexagonal jet stream at its north pole. The images captured by Cassini provided unprecedented views of these atmospheric features, allowing scientists to investigate their origins and dynamics. Porco's analysis of the data contributed to our understanding of the weather and climate on Saturn, as well as the broader atmospheric processes that occur on gas giant planets.

Throughout her career, Porco has been a passionate advocate for space exploration and science communication. She has been dedicated to sharing the excitement and wonder of space with the public, using her expertise and enthusiasm to inspire others. Porco has given numerous public lectures, participated in media interviews, and contributed to documentaries and television programs about space exploration. Her ability to convey complex scientific concepts in an engaging and accessible manner has made her a prominent and influential voice in the field.

In addition to her scientific and outreach work, Porco has been involved in various initiatives aimed at advancing planetary science and exploration. She founded and serves as the director of the nonprofit organization Diamond Sky Productions, which focuses on science education and public outreach. She has also been a vocal advocate for the continued exploration of the outer solar system, emphasizing the importance of missions to study the icy moons of Jupiter and Saturn, as well as other distant objects in the Kuiper Belt.

Porco's contributions to planetary science have been recognized with numerous awards and honors. She received the Carl Sagan Medal for Excellence in Public Communication in Planetary Science from the American Astronomical Society's Division for Planetary Sciences, the Lennart Nilsson Award for scientific photography, and the Golden Plate Award from the American Academy of Achievement, among others. Her work has been widely published in scientific journals, and she has been invited to speak at prestigious institutions and conferences around the world.

One of Porco's most significant honors came in 2010 when the International Astronomical Union named an asteroid in her honor, 7231 Porco. This recognition highlights her lasting impact on the field of planetary science and her contributions to our understanding of the solar system.

Porco's legacy extends beyond her scientific achievements and public outreach. She has served as a mentor and role model for aspiring scientists, particularly women in the field of planetary science. Her dedication to advancing knowledge and promoting scientific inquiry has inspired a new generation of researchers and explorers.

Chapter 43: Annie Jump Cannon

Annie Jump Cannon, born on December 11, 1863, in Dover, Delaware, was an American astronomer whose work was crucial in the development of contemporary stellar classification. She played a pivotal role at the Harvard College Observatory, where she devised a classification system that is still in use today. Cannon's life and career were marked by a deep dedication to astronomy, overcoming personal and societal challenges, and a lasting impact on the field of astrophysics.

Cannon was born into a well-educated family that encouraged her interests in science and astronomy from an early age. Her mother, Mary Jump Cannon, was particularly influential, fostering Annie's curiosity about the stars and teaching her the constellations. Annie's father, Wilson Cannon, was a Delaware state senator and shipbuilder who provided a stable and supportive environment for her intellectual pursuits.

Cannon's formal education began at Wellesley College in Massachusetts, where she studied physics and astronomy under the tutelage of Professor Sarah Frances Whiting. Whiting was a pioneering female physicist and an advocate for women in science, and she played a significant role in Cannon's development as a scientist. At Wellesley, Cannon graduated as valedictorian in 1884, having demonstrated a remarkable aptitude for mathematics and the sciences.

After graduation, Cannon returned home, where she faced a series of personal challenges, including the deaths of her mother and father. Despite these difficulties, her passion for astronomy remained undiminished. She spent the next decade traveling and studying photography, a skill that would later prove invaluable in her astronomical work. In 1892, Cannon contracted scarlet fever, which left her nearly deaf for the rest of her life. This disability, however, did not deter her from pursuing her scientific ambitions.

In 1894, Cannon returned to Wellesley to study advanced astronomy and physics courses. During this period, she developed a keen interest in the nascent field of spectroscopy, the study of the interaction between matter and electromagnetic radiation. Spectroscopy would become a central element of her later work in stellar classification.

Cannon's big break came in 1896 when she was hired as an assistant at the Harvard College Observatory by Edward C. Pickering, the director of the observatory. Pickering was a progressive figure who actively recruited women to work on his project, known as the "Henry Draper Catalogue," which aimed to catalog and classify stars based on their spectral characteristics. This project was funded by a bequest from the widow of Henry Draper, a pioneer in astrophotography.

At the Harvard Observatory, Cannon joined a group of women known as "Pickering's Harem" or the "Harvard Computers," who were responsible for analyzing photographic plates of stars and identifying their spectral lines. These women, including notable figures like Williamina Fleming and Antonia Maury, made significant contributions to the field of astronomy, often working in relative obscurity compared to their male counterparts.

Cannon's work at the observatory involved painstakingly examining thousands of photographic plates and identifying the spectral lines of stars. She quickly became known for her exceptional ability to recognize patterns and her meticulous attention to detail. Her work led her to develop a new system for classifying stars, known as the Harvard Classification Scheme. This system categorized stars based on their temperatures, using the spectral lines of hydrogen as a key indicator.

The Harvard Classification Scheme simplified and refined earlier classification systems, particularly those developed by her colleagues Williamina Fleming and Antonia Maury. Cannon's system arranged stars into seven main classes: O, B, A, F, G, K, and M, in order of decreasing temperature. Each class was further divided into ten subclasses, numbered 0 to 9, to provide more precise classifications. This system, often remembered by the mnemonic "Oh, Be A Fine Girl, Kiss Me," became the foundation for modern stellar classification.

Cannon's classification system was revolutionary because it provided a consistent and systematic way to categorize stars based on their spectral characteristics. Her work demonstrated that the spectra of stars could reveal important information about their temperatures, compositions, and evolutionary stages. The Harvard Classification Scheme allowed astronomers to better understand the diversity of stars in the universe and laid the groundwork for future research in astrophysics.

Cannon's contributions to astronomy extended beyond her classification system. She played a key role in compiling the Henry Draper Catalogue, which cataloged the spectral types of over 225,000 stars. This monumental work was published in multiple volumes between 1918 and 1924 and remains a valuable resource for astronomers. Cannon personally classified around 350,000 stars during her career, an incredible achievement given the manual and time-consuming nature of the work.

Despite her significant contributions, Cannon faced numerous challenges as a woman in a male-dominated field. She worked for decades in a subordinate position at the Harvard Observatory, receiving lower pay and less recognition than her male colleagues. However, her achievements gradually earned her respect and accolades from the scientific community. In 1938, she became the first woman to receive an honorary doctorate from the University of Oxford. She also received the Henry Draper Medal from the National Academy of Sciences in 1931, making her the first woman to be awarded this prestigious honor.

Cannon's legacy extends beyond her scientific achievements. She was a passionate advocate for women in science and worked to promote the careers of female astronomers. She was a founding member of the American Astronomical Society's (AAS) Committee on the Status of Women in Astronomy, which aimed to address the challenges faced by women in the field and to promote gender equality. Cannon also established the Annie Jump Cannon Award in Astronomy, which is awarded annually by the AAS to a woman who has made significant contributions to the field of astronomy.

In addition to her scientific work, Cannon was an active member of various organizations and societies. She served as the president of the American Association of Variable Star Observers (AAVSO) and was involved in the International Astronomical Union (IAU). Her dedication to the scientific community and her efforts to support the work of other astronomers were widely recognized and appreciated.

Cannon's influence on the field of astronomy continues to be felt today. Her classification system remains a fundamental tool for astronomers, and her contributions to the Henry Draper Catalogue are still used as a reference for stellar data. Her work laid the foundation for the development of the Hertzsprung-Russell diagram, which plots the luminosity of stars against their spectral types and has become a crucial tool for understanding stellar evolution.

Cannon's life and career are a testament to her determination, intelligence, and passion for astronomy. She overcame significant personal and professional challenges to make lasting contributions to the field of astrophysics. Her work not only advanced our understanding of the universe but also paved the way for future generations of female scientists.

Annie Jump Cannon passed away on April 13, 1941, at the age of 77. Her legacy endures through the continued use of her classification system and the recognition of her contributions to astronomy. She is remembered as a pioneering scientist who made significant strides in the study of stars and as a trailblazer who helped to open doors for women in science. Cannon's work and dedication serve as an inspiration to all those who seek to explore the mysteries of the cosmos and to advance the frontiers of human knowledge.

Chapter 44: Henrietta Swan Leavitt

Henrietta Swan Leavitt, born on July 4, 1868, in Lancaster, Massachusetts, was an American astronomer whose groundbreaking work on variable stars significantly advanced our understanding of the universe. Her discovery of the relationship between the luminosity and the period of Cepheid variable stars laid the foundation for measuring astronomical distances and played a crucial role in the development of modern cosmology. Despite facing numerous challenges as a woman in a male-dominated field, Leavitt's contributions to astronomy have had a lasting impact on the field.

Leavitt's early life was marked by a strong academic background and a supportive family. She was one of seven children in a well-educated family that valued learning and intellectual pursuits. Her father, George Roswell Leavitt, was a Congregational minister, and her mother, Henrietta Swan Kendrick, was a homemaker who encouraged her children to pursue education. Leavitt attended Oberlin College in Ohio before transferring to the Society for the Collegiate Instruction of Women, later known as Radcliffe College, the women's coordinate institution of Harvard University. She graduated in 1892 with a degree in astronomy.

After completing her studies, Leavitt faced a common challenge for women of her time: finding a professional position in her field. Women were largely excluded from academic and scientific careers, and opportunities for them to contribute to scientific research were limited. However, Leavitt's determination and passion for astronomy led her to the Harvard College Observatory, where she began working as a "computer" in 1893. The term "computer" referred to individuals, often women, who performed mathematical calculations and data analysis before the advent of electronic computers.

At the Harvard College Observatory, Leavitt worked under the direction of Edward Charles Pickering, who was known for hiring women to process astronomical data. These women, known as the "Harvard Computers" or "Pickering's Harem," were responsible for examining and cataloging the vast amounts of data collected by the observatory. They played a critical role in analyzing photographic plates of the night sky and identifying various celestial phenomena.

Leavitt's work primarily focused on the study of variable stars—stars that change in brightness over time. Variable stars were of particular interest to astronomers because their changing luminosity could provide valuable information about their physical properties and distances. Leavitt was assigned the task of examining variable stars in the Magellanic Clouds, two irregular dwarf galaxies that are among the closest neighbors to the Milky Way.

Through meticulous examination of the photographic plates, Leavitt discovered a significant number of variable stars in the Magellanic Clouds. She meticulously recorded their periods—the time it took for the stars to complete one cycle of brightness variation. Leavitt's keen observational skills and attention to detail allowed her to identify a pattern among these stars that would become one of the most important discoveries in astronomy.

In 1908, Leavitt published her initial findings in a paper titled "1777 Variables in the Magellanic Clouds," in which she noted a relationship between the luminosity and the period of Cepheid variable stars. She observed that brighter Cepheid variables had longer periods, while fainter ones had shorter periods. However, it was in a subsequent paper, published in 1912, where she presented her discovery more explicitly. Leavitt's work revealed a direct correlation between the period of a Cepheid variable star and its intrinsic brightness, known as the period-luminosity relation.

The period-luminosity relation was revolutionary because it provided a way to determine the absolute magnitude of Cepheid variable stars based on their periods. By comparing the absolute magnitude with the apparent magnitude observed from Earth, astronomers could calculate the distance to these stars using the inverse square law of light. This method, known as the "Leavitt Law" or "Cepheid variable method," became a crucial tool for measuring astronomical distances.

Leavitt's discovery had profound implications for the field of astronomy and cosmology. Prior to her work, measuring distances to celestial objects was a significant challenge, limiting our understanding of the scale and structure of the universe. The ability to determine distances using Cepheid variables allowed astronomers to create more accurate maps of the Milky Way and other galaxies. It also provided a means to measure the expansion of the universe and contributed to the development of the field of extragalactic astronomy.

One of the most notable applications of Leavitt's work came in the 1920s, when Edwin Hubble used Cepheid variable stars to determine the distance to the Andromeda Galaxy. Hubble's observations confirmed that Andromeda was not part of the Milky Way, as previously thought, but was a separate galaxy. This discovery fundamentally changed our understanding of the universe, revealing that it was much larger and more complex than previously imagined. Hubble's work, based on Leavitt's period-luminosity relation, laid the groundwork for the field of cosmology and the study of the large-scale structure of the universe.

Despite the significance of her contributions, Leavitt received relatively little recognition during her lifetime. She worked quietly and diligently at the Harvard Observatory, largely out of the public eye. Her male colleagues often received more credit for discoveries that were based on her work. For example, Edwin Hubble's groundbreaking discoveries were made possible by Leavitt's period-luminosity relation, yet she did not receive the same level of recognition for her role in these advancements.

Leavitt's contributions to astronomy extended beyond her work on Cepheid variables. She also made significant contributions to the study of stellar magnitudes, developing a standard system for measuring and comparing the brightness of stars. Her work in this area helped to standardize astronomical measurements and improve the accuracy of observational data.

Leavitt's career was tragically cut short by illness. In 1921, she passed away from cancer at the age of 53. Despite her relatively short life, her contributions to astronomy had a lasting impact on the field. Her work on Cepheid variables and the period-luminosity relation transformed our understanding of the universe and provided a foundation for future discoveries.

Leavitt's legacy has been increasingly recognized in the years following her death. She is now celebrated as one of the pioneers of modern astronomy, and her contributions are acknowledged in scientific literature and history. In 1925, four years after her death, she was nominated for the Nobel Prize in Physics by Gösta Mittag-Leffler, a prominent Swedish mathematician and astronomer. Although the Nobel Prize is not awarded posthumously, the nomination was a testament to the significance of her work.

In addition to the Nobel nomination, Leavitt has received numerous posthumous honors. In 1973, the crater Leavitt on the Moon was named in her honor, recognizing her contributions to astronomy. The asteroid 5383 Leavitt, discovered in 1988, was also named after her. These celestial namesakes serve as enduring reminders of her impact on the field of astronomy.

Leavitt's story is often highlighted as an example of the challenges faced by women in science and the importance of recognizing their contributions. Her perseverance and dedication to her work, despite the societal and institutional barriers she encountered, continue to inspire future generations of scientists. Leavitt's legacy is not only her scientific achievements but also her role as a trailblazer for women in astronomy and the broader scientific community.

In recent years, efforts have been made to bring greater attention to Leavitt's contributions and to ensure that her story is included in the history of science. Books, articles, and documentaries have been produced to highlight her life and work, shedding light on the often-overlooked contributions of women to the field of astronomy.

Henrietta Swan Leavitt's life and work exemplify the spirit of scientific inquiry and the pursuit of knowledge. Her discovery of the period-luminosity relation of Cepheid variable stars revolutionized our understanding of the universe and provided a crucial tool for measuring astronomical distances. Despite the challenges she faced as a woman in science, her contributions have had a lasting impact on the field of astronomy and continue to shape our understanding of the cosmos. Leavitt's legacy serves as a powerful reminder of the importance of recognizing and celebrating the contributions of all scientists, regardless of gender, and her work remains a cornerstone of modern astrophysics.

Chapter 45: Alan Stern

Alan Stern, born on November 22, 1957, in New Orleans, Louisiana, is an American planetary scientist, space program executive, and author, renowned for his pivotal role in the New Horizons mission to Pluto and the Kuiper Belt. His career has been marked by significant contributions to space exploration and planetary science, innovative leadership in space missions, and extensive public engagement and advocacy for space science.

Stern's early life was characterized by a deep fascination with space and science. His interest in astronomy and space exploration was sparked at a young age, influenced by the space race and the Apollo moon landings. Stern's passion for science led him to pursue higher education in physics and astronomy. He attended the University of Texas at Austin, where he earned a Bachelor of Science degree in Physics. He continued his education at the University of Colorado Boulder, where he obtained master's degrees in Aerospace Engineering and Planetary Atmospheres and a Ph.D. in Astrophysics and Planetary Science in 1989.

Stern's early career involved a combination of academic research and practical experience in the field of space exploration. He worked at the Johnson Space Center as a postdoctoral researcher and at the Southwest Research Institute (SwRI) in Boulder, Colorado, where he would later return as an executive. Stern's work during this period focused on the outer solar system, including the study of Pluto and the Kuiper Belt, regions that were relatively unexplored at the time.

One of Stern's most significant achievements came with his leadership of the New Horizons mission, a project that would become a landmark in space exploration. The idea for a mission to Pluto had been proposed multiple times before, but it faced numerous challenges, including funding issues, mission cancellations, and technical hurdles. Stern was a driving force behind the mission, advocating tirelessly for its approval and funding. His persistence and leadership were crucial in keeping the mission alive through various obstacles.

In 2001, NASA approved the New Horizons mission, with Stern as the Principal Investigator. The mission's goal was to conduct the first flyby of Pluto, providing detailed images and data about the distant dwarf planet and its moons. The New Horizons spacecraft was launched on January 19, 2006, embarking on a nine-and-a-half-year journey to the outer reaches of the solar system. Stern's vision and leadership were instrumental in shaping the mission's scientific objectives and ensuring its success.

The New Horizons mission reached a historic milestone on July 14, 2015, when it made its closest approach to Pluto. The spacecraft sent back stunning images and valuable scientific data, revealing a complex and diverse world with mountains of water ice, vast plains, and potential evidence of subsurface oceans. The mission's findings challenged previous assumptions about Pluto and provided new insights into the geology and atmosphere of the dwarf planet. Stern's leadership was widely recognized as a key factor in the mission's success, and he became a prominent figure in the field of planetary science.

Following the Pluto flyby, New Horizons continued its journey into the Kuiper Belt, a region of the solar system beyond Neptune that contains many small, icy bodies. On January 1, 2019, the spacecraft conducted a flyby of the Kuiper Belt object (486958) Arrokoth, previously known as Ultima Thule. This flyby marked the most distant object ever visited by a spacecraft and provided valuable data about the early solar system. The success of these extended missions further demonstrated Stern's vision and the mission team's capabilities.

In addition to his work on New Horizons, Stern has been involved in numerous other space missions and projects. He has contributed to over 24 suborbital, orbital, and planetary missions, including collaborations with NASA, the European Space Agency (ESA), and private spaceflight initiatives. His work has spanned a wide range of topics, from studying the atmospheres of planets and moons to developing instruments for space exploration.

Stern's contributions to space exploration extend beyond his technical and scientific achievements. He has been a vocal advocate for planetary science and space exploration, promoting public interest and support for these fields. Stern has authored numerous articles and books aimed at both scientific audiences and the general public, sharing the excitement and importance of space exploration. His ability to communicate complex scientific concepts in an engaging and accessible manner has made him a prominent figure in science communication.

Throughout his career, Stern has received numerous awards and honors in recognition of his contributions to planetary science and space exploration. These include the NASA Exceptional Achievement Medal, the Carl Sagan Memorial Award, and induction into the American Academy of Arts and Sciences. In 2007, Time magazine named him one of the 100 most influential people in the world, highlighting his impact on the field of space exploration.

Stern's influence extends to his role in shaping the future of space exploration. He has been involved in the development of new missions and technologies, advocating for the continued exploration of the outer solar system and beyond. His work with private spaceflight companies, including roles with Blue Origin and Virgin Galactic, reflects his commitment to advancing commercial spaceflight and expanding access to space.

In recent years, Stern has also focused on the search for habitable environments and life beyond Earth. He has been involved in projects aimed at exploring the icy moons of the outer planets, such as Europa and Enceladus, which are believed to have subsurface oceans that could potentially harbor life. Stern's advocacy for these missions highlights his forward-thinking approach and his dedication to exploring the unknown.

Stern's legacy in planetary science and space exploration is characterized by his unwavering dedication, innovative leadership, and passion for discovery. His work on the New Horizons mission has transformed our understanding of Pluto and the Kuiper Belt, providing a wealth of scientific data and inspiring future generations of scientists and explorers. Stern's advocacy for space exploration and his contributions to science communication have helped to foster a greater appreciation for the importance of exploring the cosmos.

Beyond his scientific and technical achievements, Stern's career serves as an example of the power of persistence and vision in overcoming challenges and achieving great things. His ability to navigate the complexities of space missions, secure funding, and inspire teams to reach their goals has left an indelible mark on the field of planetary science.

Chapter 46: Margaret Hamilton

Margaret Hamilton, born on August 17, 1936, in Paoli, Indiana, is an American computer scientist and systems engineer whose pioneering work in software engineering significantly contributed to the success of the Apollo missions. Her development of the onboard flight software for the Apollo spacecraft and her role in the advancement of software engineering have cemented her legacy as one of the most influential figures in the history of computing and space exploration.

Hamilton's early life was marked by an interest in mathematics and science, fields that were not typically encouraged for women during her time. She attended Hancock High School in Michigan, where she excelled in her studies. Her academic talents earned her a place at Earlham College in Richmond, Indiana, where she pursued a Bachelor of Arts degree in mathematics, graduating in 1958. At Earlham, Hamilton developed a strong foundation in mathematics, which would later prove invaluable in her career in computer science and systems engineering.

After graduating from Earlham, Hamilton moved to Boston with her husband, James Hamilton, who was studying at Harvard Law School. In Boston, she began her career as a high school mathematics teacher. However, her passion for mathematics and her desire to be involved in more challenging and innovative work led her to seek opportunities in the burgeoning field of computer science.

In the early 1960s, Hamilton took a job at the Massachusetts Institute of Technology (MIT), working in the Lincoln Laboratory. There, she was part of a team developing software for the Semi-Automatic Ground Environment (SAGE) project, an early air defense system. This experience provided her with valuable skills in programming and software development, as she worked on real-time computing and the development of complex systems.

Hamilton's work at MIT's Lincoln Laboratory eventually led her to the Instrumentation Laboratory, which later became the Charles Stark Draper Laboratory. It was here that Hamilton would make her most significant contributions to space exploration. The laboratory had been contracted by NASA to develop the onboard flight software for the Apollo missions, which aimed to land humans on the Moon and return them safely to Earth.

Hamilton joined the Apollo program in 1965 as a lead software engineer and director of the Software Engineering Division at MIT. She and her team were responsible for developing the software for the Apollo Guidance Computer (AGC), which would control the spacecraft's navigation and landing systems. This task was immensely challenging, as it required creating software that could operate reliably in the harsh and unpredictable environment of space, with limited computational resources and no room for error.

One of Hamilton's key contributions to the Apollo program was her emphasis on the importance of software reliability and robustness. She recognized that software failures could have catastrophic consequences for the mission and the astronauts' safety. To address this, Hamilton and her team developed rigorous testing and validation procedures to ensure the software's reliability. They also implemented innovative techniques for error detection and recovery, which were critical for the success of the missions.

Hamilton's work on the Apollo software was groundbreaking in many ways. She introduced the concept of "asynchronous execution," allowing the software to handle multiple tasks simultaneously, and developed priority scheduling algorithms to ensure that critical tasks were executed promptly. These innovations were essential for managing the complex and time-sensitive operations required during the Apollo missions.

One of the most famous examples of Hamilton's software's robustness occurred during the Apollo 11 mission in July 1969, the first mission to land humans on the Moon. As the Lunar Module, named Eagle, was descending towards the lunar surface, the AGC began to display a series of program alarms indicating that it was overloaded with tasks. Hamilton's software design allowed the AGC to prioritize essential tasks and safely complete the landing despite the overload. Her work ensured that the mission continued smoothly, allowing Neil Armstrong and Buzz Aldrin to make history by becoming the first humans to set foot on the Moon.

The success of the Apollo 11 mission and the subsequent Apollo missions highlighted the critical role of Hamilton's contributions to the field of software engineering. Her work demonstrated the importance of software in mission-critical systems and laid the foundation for the development of reliable and robust software for future space missions and other high-stakes applications.

In recognition of her contributions to the Apollo program and the field of software engineering, Hamilton received numerous accolades and honors. In 1986, she founded Hamilton Technologies, Inc., a company focused on the development of advanced software methodologies. Her work continued to influence the field of software engineering, and she became an advocate for the importance of rigorous software development practices.

One of the most significant recognitions of Hamilton's contributions came in 2016 when she was awarded the Presidential Medal of Freedom, the highest civilian honor in the United States, by President Barack Obama. This honor acknowledged her pioneering work in software engineering and her crucial role in the success of the Apollo missions. During the award ceremony, President Obama highlighted Hamilton's contributions, stating that her work helped enable the Moon landing and inspired future generations of scientists and engineers.

Hamilton's influence extends beyond her technical achievements. As a woman working in a male-dominated field, she broke barriers and paved the way for future generations of women in science, technology, engineering, and mathematics (STEM). Her success demonstrated that women could excel in these fields and make significant contributions to science and technology. Hamilton became a role model and mentor for many aspiring female scientists and engineers, advocating for greater inclusion and diversity in STEM disciplines.

Throughout her career, Hamilton has been a vocal advocate for the importance of software engineering and its role in modern technology. She coined the term "software engineering" to describe the systematic and disciplined approach needed to develop reliable and robust software systems. Her emphasis on the engineering aspects of software development helped to elevate the field and establish it as a critical discipline within the broader field of engineering.

Hamilton's legacy continues to inspire and influence the field of software engineering. Her work on the Apollo missions remains a testament to the importance of rigorous software development practices and the impact of software on mission-critical systems. The techniques and methodologies she developed for the Apollo software have been adapted and applied to a wide range of applications, from aerospace and defense to healthcare and financial systems.

In addition to her technical contributions, Hamilton has been an advocate for the importance of interdisciplinary collaboration in solving complex problems. She recognized that successful software development requires collaboration between engineers, scientists, and domain experts. Her approach to software engineering emphasized the need for clear communication, teamwork, and a deep understanding of the problem domain.

Hamilton's contributions to the field of software engineering and her role in the Apollo missions have been the subject of numerous books, articles, and documentaries. Her story has been highlighted as an example of the critical role of software in space exploration and the importance of recognizing the contributions of women in science and engineering. Hamilton's work has also been featured in educational programs and exhibits, inspiring future generations of scientists and engineers.

Chapter 47: Mae Jemison

Mae Jemison, an American engineer, physician, and former NASA astronaut, made history as the first African-American woman to travel in space. Born on October 17, 1956, in Decatur, Alabama, Jemison's journey to becoming a trailblazing astronaut is marked by extraordinary accomplishments, a deep passion for science and education, and a relentless commitment to advocating for diversity and inclusion in STEM (Science, Technology, Engineering, and Mathematics).

Jemison was the youngest of three children born to Charlie Jemison, a maintenance supervisor, and Dorothy Jemison, an elementary school teacher. When she was three years old, her family moved to Chicago, Illinois, in search of better educational and employment opportunities. Growing up in Chicago, Jemison was nurtured by a supportive family environment that emphasized the importance of education and hard work. Her parents encouraged her curiosity and fostered her interests in science and the arts.

From a young age, Jemison showed an aptitude for science. She was particularly inspired by the Apollo missions and the achievements of the astronauts, despite not seeing any African-Americans or women among them. Jemison's interest in space was further fueled by the television show "Star Trek," especially the character of Lieutenant Uhura, portrayed by Nichelle Nichols. This representation of a black woman in a significant role in a science fiction series left a lasting impact on her and broadened her horizons regarding what was possible.

Jemison's academic journey was marked by excellence. She attended Morgan Park High School, where she was an honor student and actively involved in extracurricular activities, including the science club and dance. After graduating in 1973, Jemison entered Stanford University at the age of 16. She pursued a double major in Chemical

Engineering and African and African-American Studies, reflecting her diverse interests and commitment to both science and social issues. At Stanford, Jemison faced challenges related to being a young African-American woman in a predominantly white, male environment. However, she remained undeterred and excelled academically while also engaging in activities such as dance and theater.

In 1977, Jemison graduated from Stanford with a Bachelor of Science degree in Chemical Engineering and a Bachelor of Arts degree in African and African-American Studies. She then enrolled at Cornell University Medical College, where she continued to break barriers and pursue her passion for helping others. During her medical training, Jemison traveled to Cuba, Kenya, and Thailand to provide medical care to underserved populations, gaining valuable experience in international health and medical practice in diverse environments. She also worked as a volunteer in a Cambodian refugee camp, demonstrating her commitment to humanitarian efforts.

After earning her Doctor of Medicine (M.D.) degree from Cornell in 1981, Jemison completed her medical internship at Los Angeles County+USC Medical Center and later worked as a general practitioner. However, her ambitions extended beyond the confines of traditional medical practice. Driven by a desire to explore new frontiers and make a broader impact, Jemison joined the Peace Corps in 1983. She served as a medical officer in Liberia and Sierra Leone for two years, where she managed the healthcare delivery system for Peace Corps volunteers and worked with the local health systems to improve medical care. Her experiences in the Peace Corps reinforced her belief in the interconnectedness of global health, technology, and social development.

Jemison's journey took a pivotal turn when she decided to pursue her childhood dream of becoming an astronaut. In 1987, she applied to NASA's astronaut program. Her application came at a time when NASA was looking to diversify its astronaut corps and encourage women and minorities to apply. Out of thousands of applicants, Jemison was one of the 15 candidates selected for NASA Astronaut Group 12, the first group chosen after the Challenger disaster in 1986. Her selection made her the first African-American woman admitted into NASA's astronaut training program.

Jemison's training included intensive preparation in areas such as space science, flight training, survival techniques, and technical skills necessary for space missions. She completed her training in 1988 and was assigned to various technical tasks within NASA, including work on the Space Shuttle's software verification and payload bay. Her training and assignments equipped her with the knowledge and skills needed for her upcoming mission.

On September 12, 1992, Jemison launched into space aboard the Space Shuttle Endeavour on mission STS-47. The mission was a cooperative effort between the United States and Japan and focused on life and material sciences experiments in space. Jemison served as a Mission Specialist, and her responsibilities included conducting experiments on weightlessness and motion sickness, as well as other scientific investigations. During the eight-day mission, Jemison and her crewmates conducted over 40 experiments, contributing valuable data to the understanding of space travel's effects on the human body and other scientific fields.

Jemison's historic journey into space was a significant milestone, not only for her personal achievements but also for its symbolic importance. Her presence aboard the Endeavour challenged stereotypes and broke down barriers, inspiring countless individuals, especially women and people of color, to pursue careers in STEM fields. Jemison's accomplishment demonstrated that space exploration was an inclusive endeavor, open to all who had the passion, dedication, and qualifications.

After her return from space, Jemison resigned from NASA in 1993 to pursue other interests and focus on her broader vision for science and technology. She founded The Jemison Group, a technology consulting firm that focuses on the integration of science and socio-cultural issues. Through her company, Jemison has worked on projects that aim to improve healthcare, environmental sustainability, and educational opportunities, particularly in developing countries.

In 1994, Jemison founded the Dorothy Jemison Foundation for Excellence, named in honor of her mother. The foundation focuses on promoting science literacy and education, with a particular emphasis on underserved communities. One of its flagship programs is The Earth We Share (TEWS), an international science camp for students aged 12 to 16. TEWS encourages students to solve global problems through science and technology, fostering critical thinking, teamwork, and leadership skills. Through her foundation and educational initiatives, Jemison has continued to inspire and empower young people to pursue careers in STEM and make meaningful contributions to society.

Jemison has also been a prominent advocate for increasing diversity in STEM fields. She has spoken extensively about the importance of representation and the need for more women and minorities in science and technology. Her advocacy efforts have included participating in panels, delivering keynote addresses, and collaborating with organizations dedicated to promoting diversity and inclusion. Jemison's work has helped to raise awareness about the barriers faced by underrepresented groups and the importance of creating inclusive environments where everyone can thrive.

In addition to her work in education and advocacy, Jemison has held academic positions and contributed to various scientific and technological endeavors. She has served as a professor of environmental studies at Dartmouth College and has been a professor-at-large at Cornell University. Jemison has also been involved in initiatives such as the 100 Year Starship project, an ambitious effort to ensure that human space travel beyond our solar system is possible within the next century. Her involvement in this project reflects her continued passion for exploration and her belief in the boundless possibilities of human ingenuity.

Throughout her career, Jemison has received numerous awards and honors in recognition of her contributions to science, education, and space exploration. These accolades include induction into the National Women's Hall of Fame and the International Space Hall of Fame, as well as honorary doctorates from several universities. Jemison's achievements have been celebrated in various media, including books, documentaries, and exhibitions, highlighting her enduring impact on society.

Mae Jemison's legacy is one of breaking barriers, challenging stereotypes, and inspiring future generations. Her journey from a young girl with dreams of space to a pioneering astronaut and influential advocate exemplifies the power of perseverance, education, and vision. Jemison's contributions to science and technology have opened doors for countless individuals and have helped to create a more inclusive and equitable world.

Jemison continues to be an influential figure in science and education. She remains active in promoting STEM education, advocating for diversity, and working on projects that address global challenges. Her dedication to fostering a love of science and exploration in young people, particularly those from underserved communities, ensures that her impact will be felt for generations to come.

Chapter 48: Sally Ride

Sally Ride, an American astronaut and physicist, holds a significant place in history as the first American woman to travel into space. Born on May 26, 1951, in Los Angeles, California, Ride's journey from a curious young girl with a passion for science and tennis to a pioneering astronaut and dedicated advocate for science education is a story of groundbreaking achievements, perseverance, and a commitment to inspiring future generations.

Sally Kristen Ride grew up in Encino, a suburb of Los Angeles, where her parents, Dale Burdell Ride and Carol Joyce Ride, nurtured her curiosity and intellectual pursuits. Her father was a political science professor, and her mother was a counselor, providing a supportive environment that valued education and exploration. From an early age, Ride demonstrated a keen interest in science, especially physics, and a remarkable talent for tennis, competing at a national level during her teenage years.

Ride attended Westlake School for Girls, a private preparatory school in Los Angeles, where she excelled academically and athletically. Her diverse interests and achievements in both academics and sports earned her a scholarship to Swarthmore College. However, she left Swarthmore after three semesters to focus on her tennis career, aspiring to become a professional player. Despite her dedication, she eventually realized that her true passion lay in science rather than sports.

In 1970, Ride enrolled at Stanford University, where she pursued a double major in physics and English. At Stanford, she found an environment that allowed her to fully immerse herself in her scientific studies while continuing to play competitive tennis. Ride's academic excellence and research potential were quickly recognized, and she completed her Bachelor of Science in Physics and Bachelor of Arts in English in 1973. She continued her studies at Stanford, earning a Master of Science degree in 1975 and a Ph.D. in Physics in 1978, focusing on astrophysics and free electron lasers.

Ride's path to becoming an astronaut began with a newspaper advertisement. In 1977, NASA began recruiting for its new class of astronauts, the first to include women and minorities. Ride responded to the advertisement and was selected from a pool of 8,000 applicants, becoming one of six women accepted into NASA's astronaut program. This group, known as NASA Astronaut Group 8, was a historic cohort that marked a significant shift towards inclusivity and diversity in the space program.

Ride's astronaut training included intensive preparation in various areas, such as spacecraft systems, survival training, and flight simulations. Her scientific background, particularly her expertise in physics, made her an ideal candidate for missions that required a deep understanding of technical and scientific principles. Ride's dedication and performance during training set her apart, and she was soon assigned to mission specialist roles for upcoming Space Shuttle missions.

Ride's first spaceflight was on the Space Shuttle Challenger during mission STS-7, which launched on June 18, 1983. As a mission specialist, Ride played a crucial role in the deployment of two communications satellites and conducted experiments using the shuttle's robotic arm, the Remote Manipulator System (RMS). Her presence on the flight made her the first American woman in space, a milestone that garnered significant media attention and public interest. Ride's composure, professionalism, and technical proficiency during the mission highlighted her capabilities as an astronaut and a role model.

The STS-7 mission was a success, and Ride's performance earned her a place on a second flight. In 1984, she flew again on the Challenger for mission STS-41-G, which included the first crew of seven astronauts and the first mission to carry two women, with Ride joined by astronaut Kathryn Sullivan. During this mission, Ride continued to demonstrate her expertise with the RMS and contributed to various scientific and technological experiments, including the first spacewalk by an American woman, performed by Sullivan.

Ride was scheduled for a third flight, STS-61-M, which was ultimately canceled due to the Challenger disaster on January 28, 1986. The tragedy profoundly affected Ride and the entire NASA community, leading to a hiatus in the shuttle program and extensive investigations into the causes of the accident. Ride was appointed to the Rogers Commission, the presidential commission tasked with investigating the disaster. Her contributions to the commission were vital in understanding the technical and organizational failures that led to the tragedy and in recommending improvements to enhance the safety and reliability of future missions.

Following her work on the Rogers Commission, Ride continued to serve NASA in various capacities. She was assigned to NASA Headquarters in Washington, D.C., where she led strategic planning and worked on long-term space exploration initiatives. Ride's contributions during this period included developing NASA's strategic goals and promoting scientific research and international cooperation in space exploration.

In 1987, Ride left NASA to pursue academic and educational endeavors. She became a professor of physics at the University of California, San Diego (UCSD), where she also served as the Director of the California Space Institute. In these roles, Ride focused on research and education, continuing to contribute to the scientific community while inspiring and mentoring students. Her academic work included research in nonlinear optics and the interaction of electromagnetic waves with matter, building on her expertise in physics and astrophysics.

Ride's passion for education and outreach led her to co-author several science books for children, aimed at inspiring young readers to explore science and space. These books, co-written with her life partner, Tam O'Shaughnessy, included titles such as "To Space and Back" (1986), "Voyager: An Adventure to the Edge of the Solar System" (1992), and "The Third Planet: Exploring the Earth from Space" (2004). Through her writing, Ride sought to make science accessible and engaging, encouraging children to pursue their interests in STEM fields.

In 2001, Ride co-founded Sally Ride Science, an organization dedicated to promoting STEM education, particularly for girls and underrepresented minorities. The organization developed innovative programs and resources designed to ignite students' interest in science and technology. Sally Ride Science's initiatives included science festivals, teacher training, and online platforms that provided interactive science content and career exploration tools. Ride's vision for the organization was to create a lasting impact by inspiring the next generation of scientists, engineers, and innovators.

Ride's contributions to science, education, and space exploration have been widely recognized and honored. She received numerous awards and accolades, including the National Space Society's von Braun Award, the Lindbergh Eagle, and induction into the Astronaut Hall of Fame. In 2013, Ride was posthumously awarded the Presidential Medal of Freedom, the highest civilian honor in the United States, in recognition of her pioneering achievements and her efforts to advance science and education.

Throughout her life, Ride remained a private person, often shying away from the public spotlight despite her historic accomplishments. She was known for her modesty, intellectual rigor, and commitment to excellence. Ride's legacy extends beyond her achievements in space; she is remembered as a trailblazer who broke barriers and paved the way for future generations of women and minorities in STEM fields.

Ride's impact on the world of science and space exploration continues to be felt today. Her work with Sally Ride Science and her advocacy for STEM education have inspired countless young people to pursue careers in science and technology. The organization's programs and resources continue to engage students and educators, fostering a love of learning and exploration.

Chapter 49: Jocelyn Bell Burnell

Jocelyn Bell Burnell, a distinguished astrophysicist, is best known for her discovery of pulsars, one of the most significant astronomical findings of the 20th century. Born on July 15, 1943, in Lurgan, Northern Ireland, Burnell's journey from a curious young girl with a passion for science to an internationally renowned scientist is marked by perseverance, groundbreaking achievements, and a profound impact on the scientific community.

Susan Jocelyn Bell, as she was born, was raised in a supportive family that valued education and intellectual curiosity. Her father, G. Philip Bell, was an architect who played a role in the design of the Armagh Planetarium, and her mother, M. Allison Bell, was a stay-at-home parent who encouraged her daughter's interests. The family's frequent visits to the planetarium fostered Bell's early fascination with astronomy. However, her academic journey was not without challenges. At the age of 11, she failed an important exam that determined her educational path, leading her to attend a Quaker boarding school, where she found a nurturing environment that promoted independent thinking and scientific inquiry.

Bell's academic excellence continued to flourish at The Mount School in York, England, where she was encouraged to pursue her interest in science. She later attended the University of Glasgow, earning a Bachelor of Science degree in Physics in 1965. Her time at Glasgow was pivotal, as it laid the foundation for her future research and introduced her to the field of radio astronomy. Bell then moved to the University of Cambridge to pursue a Ph.D. in radio astronomy under the supervision of Antony Hewish, a prominent radio astronomer.

It was during her Ph.D. research at Cambridge that Bell made her landmark discovery. In 1967, as a postgraduate student, she was part of a team constructing a large radio telescope to study quasars, which are distant and highly luminous astronomical objects powered by black holes. Bell was tasked with analyzing the vast amounts of data generated by the telescope, a labor-intensive process that involved examining miles of paper chart recordings. In the midst of this meticulous work, she noticed a series of unusual and regular signals that did not fit the expected patterns of known astronomical sources.

These signals, which repeated with remarkable regularity every 1.337 seconds, initially puzzled Bell and her colleagues. After ruling out potential sources of interference and other celestial objects, Bell and Hewish concluded that they had discovered a new type of astronomical phenomenon. They dubbed the source of the signals "pulsars," short for "pulsating radio stars." Pulsars are rapidly rotating neutron stars that emit beams of electromagnetic radiation from their magnetic poles. As the star spins, these beams sweep across space, and if aligned with Earth, they are detected as regular pulses of radio waves.

The discovery of pulsars was announced in 1968 and immediately revolutionized the field of astronomy. Pulsars provided a new tool for understanding the fundamental properties of neutron stars and offered insights into the behavior of matter under extreme conditions. The implications of this discovery extended beyond astrophysics, influencing other areas of physics and contributing to the study of gravitational waves, general relativity, and the interstellar medium.

Despite her crucial role in the discovery, Bell's contributions were initially overshadowed by her supervisor, Antony Hewish, and their colleague Martin Ryle, who received the 1974 Nobel Prize in Physics for the discovery of pulsars. The omission of Bell from the Nobel Prize sparked controversy and highlighted the gender biases prevalent in the scientific community. Many prominent scientists and commentators have since argued that Bell deserved to share in the Nobel recognition. Bell herself has consistently handled the situation with grace, emphasizing the collaborative nature of scientific research and continuing her influential work in the field.

After completing her Ph.D. in 1969, Bell Burnell embarked on a distinguished academic career, holding various research and teaching positions at universities across the United Kingdom. Her roles included positions at the University of Southampton, University College London, and the Royal Observatory in Edinburgh. In 1982, she became a professor of physics at the Open University, where she played a significant role in developing the university's science curriculum and promoting distance education.

Bell Burnell's research interests extended beyond pulsars, encompassing various aspects of astrophysics, including gamma-ray bursts, X-ray binaries, and the structure of our galaxy. Her work has contributed to a deeper understanding of the cosmos and has inspired numerous students and researchers to pursue careers in astronomy and physics. Throughout her career, Bell Burnell has been a passionate advocate for increasing the participation of women and underrepresented groups in science. She has actively worked to break down barriers and promote diversity within the scientific community.

In addition to her academic and research contributions, Bell Burnell has held several leadership roles in professional organizations. She served as president of the Royal Astronomical Society from 2002 to 2004 and was the first female president of the Institute of Physics from 2008 to 2010. Her leadership in these organizations has helped to shape the direction of astrophysical research and promote the importance of science education and outreach.

Bell Burnell's achievements have been recognized with numerous awards and honors. She was made a Commander of the Order of the British Empire (CBE) in 1999 and was elevated to Dame Commander of the Order of the British Empire (DBE) in 2007. In 2018, she was awarded the Special Breakthrough Prize in Fundamental Physics, a prestigious award that included a $3 million prize. Demonstrating her commitment to fostering diversity and supporting the next generation of scientists, Bell Burnell donated the entire prize to fund scholarships for women, underrepresented minorities, and refugees pursuing careers in physics.

Throughout her life, Bell Burnell has remained deeply committed to science communication and public engagement. She has given numerous public lectures, written popular science articles, and participated in media interviews, sharing her passion for astronomy and her experiences as a pioneering scientist. Her ability to convey complex scientific concepts in an accessible and engaging manner has made her a beloved figure in the scientific community and beyond.

Bell Burnell's impact on the field of astrophysics and her contributions to promoting diversity in science have left a lasting legacy. Her discovery of pulsars opened up new avenues of research and provided valuable insights into the fundamental properties of the universe. Her advocacy for women and underrepresented groups in science has helped to create a more inclusive and equitable scientific community, inspiring countless individuals to pursue careers in STEM fields.

In reflecting on her career and achievements, Bell Burnell has often emphasized the importance of perseverance, curiosity, and the collaborative nature of scientific research. Her story is a testament to the power of passion and dedication in overcoming obstacles and achieving groundbreaking discoveries. As a role model and mentor, Bell Burnell has inspired generations of scientists to follow in her footsteps and continue exploring the mysteries of the universe.

Jocelyn Bell Burnell's contributions to science extend far beyond her discovery of pulsars. Her work has had a profound impact on our understanding of the cosmos and has paved the way for future discoveries in astrophysics. Her dedication to promoting diversity and inclusion in science has helped to create a more equitable and inclusive scientific community. Bell Burnell's legacy is one of groundbreaking achievements, unwavering commitment to science, and a passion for inspiring the next generation of scientists. Her story serves as an enduring inspiration to all who seek to explore the wonders of the universe and make meaningful contributions to the world of science.

215

Chapter 50: Viktor Safronov

Viktor Sergeyevich Safronov was a distinguished Russian astronomer and astrophysicist whose groundbreaking work on the formation of the solar system has had a profound and lasting impact on the field. Born on October 11, 1917, in Gavrilovo, Russia, Safronov's scientific career was characterized by his innovative theories and meticulous research, which laid the foundation for modern understanding of planetary formation. His contributions, particularly the Safronov model of planetesimal accretion, revolutionized the study of cosmogony and continue to influence contemporary research in planetary science.

Safronov was born into a period of significant political and social upheaval, but he developed a keen interest in science from an early age. His academic journey began at Moscow State University, where he pursued studies in physics and astronomy. Safronov's early work focused on the dynamics of celestial bodies, an interest that would guide his career and lead to some of his most significant contributions to astrophysics.

In the mid-20th century, the prevailing theories of planetary formation were incomplete and lacked a comprehensive framework. Safronov recognized the need for a more detailed and robust model to explain the processes that lead to the formation of planets and other celestial bodies. His approach combined theoretical physics with observational data, creating a holistic understanding of planetary formation.

Safronov's most notable contribution to astrophysics is the planetesimal hypothesis, which he first proposed in the 1960s. According to this hypothesis, planets form from the gradual accumulation of small solid particles, or planetesimals, within a protoplanetary disk surrounding a young star. These planetesimals collide and coalesce over time, growing into larger bodies and eventually forming planets. This model provided a comprehensive explanation for the observed distribution and characteristics of planets within the solar system.

The planetesimal hypothesis marked a significant departure from previous theories, which often failed to account for the complex interactions and processes involved in planetary formation. Safronov's work introduced key concepts such as the coagulation of dust grains, the role of gravity in accretion, and the differentiation of planetary bodies based on their composition and distance from the central star. These ideas were detailed in his seminal work, "Evolution of the Protoplanetary Cloud and Formation of the Earth and Planets," published in 1969.

In his book, Safronov meticulously described the stages of planetary formation, from the initial condensation of gas and dust in the protoplanetary disk to the formation of solid planetesimals and their subsequent growth into planets. He introduced mathematical models to describe the dynamics of particle collisions and accretion, providing a quantitative framework for understanding planetary formation. Safronov's equations and models became foundational tools for researchers studying the early stages of planetary systems.

One of Safronov's key insights was the recognition of the role of angular momentum in shaping the structure and evolution of the protoplanetary disk. He demonstrated that the distribution of angular momentum within the disk influences the formation of different types of planets, with rocky terrestrial planets forming closer to the star and gas giants forming further out. This concept helped to explain the observed arrangement of planets in the solar system and provided a framework for studying exoplanetary systems.

Safronov's work also highlighted the importance of collisional processes in planetary formation. He showed that the growth of planetesimals is governed by a balance between accretion and fragmentation, with larger bodies growing through the accumulation of smaller particles while also being subject to destructive collisions.

This insight was crucial for understanding the size distribution of objects in the asteroid belt and the formation of planetary satellites.

In addition to his theoretical work, Safronov made significant contributions to the study of planetary atmospheres and the early history of the Earth. He investigated the processes of outgassing and atmospheric loss, providing insights into the evolution of planetary atmospheres and their impact on surface conditions. Safronov's research on the early Earth shed light on the formation of the Moon, the role of giant impacts in shaping the planet's structure, and the conditions necessary for the emergence of life.

Safronov's pioneering work earned him widespread recognition and numerous accolades within the scientific community. He was a member of the Soviet Academy of Sciences and received several prestigious awards for his contributions to astrophysics. Despite the political and cultural barriers of his time, Safronov's research reached an international audience and influenced scientists around the world.

One of the most significant aspects of Safronov's legacy is the enduring relevance of his models and theories. The Safronov model of planetesimal accretion remains a cornerstone of planetary science, and subsequent research has built upon his ideas to explore new frontiers in the study of planetary systems. Modern computational simulations and observational data from space missions have confirmed many of Safronov's predictions, demonstrating the robustness and accuracy of his models.

Safronov's work has also inspired new generations of scientists to investigate the origins and evolution of planetary systems. His approach to combining theoretical models with observational data has become a standard methodology in the field, guiding research on topics such as the formation of exoplanets, the dynamics of debris disks, and the processes that govern the early evolution of planetary atmospheres.

In addition to his scientific contributions, Safronov was known for his dedication to mentoring young researchers and fostering collaboration within the scientific community. He played a key role in establishing research programs and institutions focused on planetary science, and his mentorship helped to train many of the leading astronomers and astrophysicists of the next generation.

Safronov's impact on the field of planetary science extends beyond his lifetime. His models and theories continue to be a foundational framework for understanding the processes that shape planetary systems, and his legacy is reflected in the ongoing research and discoveries in the field. The concepts and methods he developed have become integral to the study of planetary formation, guiding scientists as they explore the complexities of the cosmos.

Epilogue

As we bring our journey through "Profiles of Space Scientists: Pioneering Minds in Cosmic Exploration" to a close, we reflect on the extraordinary tapestry woven by the lives and achievements of these remarkable individuals. From the groundbreaking discoveries of Galileo and Newton to the pioneering space missions of Armstrong and Tereshkova, the scientists and explorers chronicled in this book have collectively propelled humanity's understanding of the cosmos to unprecedented heights.

The story of space exploration is far from complete. The universe, vast and enigmatic, still holds countless mysteries waiting to be uncovered. The work of these fifty scientists has laid a foundation upon which future generations will build, continuing the quest to understand our place in the cosmos. Their legacies are not just historical footnotes but living inspirations that drive contemporary research and exploration.

In this age of rapid technological advancement and international collaboration, the spirit of discovery thrives more than ever. Missions to Mars, the exploration of distant asteroids, the study of exoplanets, and the search for extraterrestrial life are but a few of the exciting frontiers we are now poised to explore. Today's scientists and engineers, standing on the shoulders of giants, are pushing the boundaries of what we know and what we can achieve.

The stories of these pioneering minds also remind us of the diverse paths that lead to great discoveries. They came from different backgrounds, faced unique challenges, and brought their own perspectives to their work. This diversity has been a strength, driving innovation and broadening our understanding. It is a testament to the idea that anyone, regardless of their origin or circumstances, can contribute to the grand narrative of space exploration.

As we look ahead, it is essential to foster this diversity and encourage the next generation of scientists, engineers, and explorers. Education and inspiration are key. By sharing the stories of those who have come before, we can ignite the imaginations of young minds and equip them with the knowledge and skills they need to make their own contributions to space science.

The universe remains a boundless frontier, and our exploration of it is a continuous, collective endeavor. The pioneering minds featured in this book have shown us that curiosity, perseverance, and creativity are the cornerstones of discovery. They have demonstrated that the pursuit of knowledge is a noble and unending quest, one that transcends borders and generations.

In the spirit of these great scientists, let us continue to explore, to question, and to reach for the stars. The cosmos awaits, and with each new discovery, we move one step closer to understanding the grand tapestry of existence. The journey is long and filled with challenges, but as the lives of these trailblazers have shown, it is a journey well worth undertaking.

May the stories within these pages inspire you to dream big, to pursue knowledge with passion, and to contribute your own chapter to the ongoing saga of human exploration. The stars are not just above us; they are within us, guiding us forward as we seek to unravel the mysteries of the universe.

The End.

www.ingramcontent.com/pod-product-compliance
Lightning Source LLC
Chambersburg PA
CBHW051425130726

47987CB00005B/1924